Peter Nijkamp
Aura Reggiani (Eds.)

Nonlinear Evolution of Spatial Economic Systems

With 75 Figures

Springer-Verlag

Berlin Heidelberg New York
London Paris Tokyo
Hong Kong Barcelona
Budapest

Prof. Dr. PETER NIJKAMP
Free University
Department of Economics
De Boelelaan 1105
1081 HV Amsterdam
The Netherlands

Prof. Dr. AURA REGGIANI
University of Bologna
Department of Economics
Piazza Scaravilli, 2
40126 Bologna
Italy

ISBN-13:978-3-642-78465-1   e-ISBN-13:978-3-642-78463-7
DOI: 10.1007/978-3-642-78463-7

This work is subject to copyright. All rights are reserved, whether the whole or part of the material is concerned, specifically the rights of translation, reprinting, reuse of illustrations, recitation, broadcasting, reproduction on microfilms or in other ways, and storage in data banks. Duplication of this publication or parts thereof is only permitted under the provisions of the German Copyright Law of September 9, 1965, in its version of June 24, 1985, and a copyright fee must always be paid. Violations fall under the prosecution act of the German Copyright Law.

© Springer-Verlag Berlin · Heidelberg 1993
Softcover reprint of the hardcover 1st edition 1993

The use of registered names, trademarks, etc. in this publication does not imply, even in the absence of a specific statement, that such names are exempt from the relevant protective laws and regulations and therefore free for general use.

42/7130-5 4 3 2 1 0 - Printed on acid-free paper

Dedication

*To all our descendants who believe that nonlinear evolution
contributes to the beauty of our world.*

PREFACE

Is our world more dynamic than it used to be in the past? Have phenomena in the social science field become unpredictable? Are chaotic events nowadays occurring more frequently than in the past? Such questions are often raised in popular debates on nonlinear evolution and self-organizing systems.

At the same time, many scientists are also raising various intruiging methodological issues. Is it possible to separate deterministic chaos from random disturbances if their trajectories are (almost) similar? Is prediction still possible in a world of chaos (Poincaré)? Is it possible to distinguish specification errors from measurement errors in a nonlinear dynamic model? Is evolution a random process?

The list of such questions can easily be extended with dozens of others. But despite the myriad of questions on problems of nonlinear evolution, one common trait is evident: in both the natural and the social sciences we are still groping in the dark in areas which are *par excellence* promising hunting grounds for exploratory and exploratory research, viz. structural grounds in an uncertain nonlinear world.

The present book aims at offering a collection of refreshing contributions to the above research issues by focusing attention, in particular on nonlinear dynamic evolution in space at the Netherlands Institute for Advanced Study (NIAS) in Wassenaar, the Netherlands. The Institute has to be thanked for its hospitality and support, reflected *inter alia* in a workshop at which several of the papers included in this book were discussed.

Particular thanks go to Dianne Biederberg (Contact Europe) who was responsible for the editing of this book, and to the Italian Consiglio Nazionale delle Ricerche (C.N.R.) which also supported the present initiative.

Amsterdam - Milano - Amsterdam, flight KL 341-342.

Peter Nijkamp
Aura Reggiani

CONTENTS

	Page
Preface	vii

1. Nonlinear Evolution of Socioeconomic and Spatial Systems — Peter Nijkamp and Aura Reggiani — 1

PART A: NONLINEAR DYNAMICS IN ECONOMICS — 23

2. Lessons from Nonlinear Dynamics Economics — Peter Nijkamp and Jacques Poot — 25

3. Survey of Nonlinear Dynamic Modelling in Economics — John Barkley Rosser, Jr. — 58

4. Towards a Dynamic Disequilibrium Theory of Economy — Gunther Haag, M. Hilliges and K. Teichmann — 85

5. Complex Transient Motion in Continuous-Time Economic Models — Hans-Walter Lorenz — 112

6. Economic Structure and Nonlinear Dynamic Development — Wei-Bin Zhang — 138

PART B: SPATIAL REPRESENTATIONS OF NONLINEAR DYNAMIC SYSTEMS — 157

7. Microeconomics and the Dynamic Modelling of Spatial Systems — William D. Macmillan — 159

8. Spatial Evolution in Complex Systems — Peter M. Allen — 178

9. Speculations on Fractal Geometry in Spatial Dynamics — Paul Longley and Michael Batty — 203

10.	Complex Dynamics and Fractal Urban Form Roger White and Guy Engelen	223
11.	Complex Behaviour in Spatial Networks Peter Nijkamp and Aura Reggiani	248
12.	Dynamical Basis of Large Deviations and Power Laws in Complex Systems Gregoire Nicolis	272
Contributors		283

CHAPTER 1
NONLINEAR EVOLUTION OF SOCIOECONOMIC AND SPATIAL SYSTEMS

Peter Nijkamp and Aura Reggiani

1.1 PROLOGUE

"Evolution, or order of process, is more than just a paradigm for the biological domain, it is a view of how a totality that hangs together in all of its interactive processes moves" (Jantsch, 1976, p. 9). The evolutionary paradigm is recently coming to the fore in modern economic and social sciences as a reaction against conventional theories aiming to reduce many complex phenomena to an equilibrium projection, system predictability and hierarchical control. In this context, the present volume aims to offer new insights into evolving complex socioeconomic modelling approaches by bringing together various advances and recent methodologies - emerging from different disciplines but applicable to dynamical spatial economics - which seem capable of improving our understanding and our design of structural (e.g. 'nested') principles of socioeconomic life such as (non)-equilibrium, (non)-predictability, (un)certainty, adaptiveness and complexity.

A fundamental issue in recent scientific debates on evolutionary principles concerns the dynamic modelling aspect which - by using nonlinear analysis - is an essential characteristic of this new research field. Starting from the relevance of nonlinear modelling, with a particular emphasis on chaos theory, we will discuss in the next sections various important issues concerning the evolution of the space-economy in light of recent achievements in ecologically-based theories.

1.2 TOWARDS NONLINEAR MODELLING

Despite the strong modelling traditions in various social sciences (inter alia economics, geography, regional science, transportation science), surprisingly little attention has been given to **nonlinear** models. Most scientists went to great length in suggesting and motivating linearized procedures as solution strategies for nonlinear models. Admittedly, linear dynamic models are certainly able to generate **unstable** solutions, but the solutions of such models are restricted to the following four regular standard types: oscillatory and stable, oscillatory and explosive, monotonous changing and stable, and monotonous changing and explosive. Such models may provide approximate replications of short-or medium-term changes, but fail to encapsulate long-term developments characterized by structure shifts of an irregular or abrupt nature. An interesting example of the increasing interest in such long-range development issues can be found in an article by Day and Walter (1990) describing the socioeconomic historical evolution of societies on the basis of 'regime switches' in technology, demography and institutions; a further example in this context is given in an article by Dendrinos and Rosser (1992), where the foundations for a comprehensive dynamic theory of discontinuities in urban population size is discussed.

The interest in modelling long-run future development paths dates back to the late sixties/early seventies with the emergence of **dynamic systems theory** (see among others Forrester, 1968, and Meadows and Meadows, 1973). Such models - mainly simulation models - focused attention on the impact of feedback effects in dynamic systems on stable behaviour. The availability of computer software led to a high degree of popularity of such experiments, although also severe criticism arose. In contrast to linear models, such systems dynamic models were able to generate also a-periodic growth patterns.

The interest in unstable systems behaviour remained in the 1970s and 1980s, and generated new contributions in the field of structure changes (such as tensor analysis, singularity theory, bifurcation theory, etc.). Especially **catastrophe theory**[1] gained much popularity in the late seventies, mainly because it is able to describe turning points and jumps - of an asymmetric and sometimes irreversible nature - in nonlinear systems (see for various illustrations Poston and Stewart, 1978, and Zeeman, 1977). In such systems the direction of movement and the level of threshold values determines the occurrence of catastrophes, which are essentially characterized by the fact that a dynamic system may have various equilibrium points for the same values of state variables.

Applications of catastrophe theory in the field of regional science can amongst others be found in Casti (1979), De Palma and Lefèvre (1987), Fischer and Jammernegg (1986), Lombardo and Rabino (1983), Nijkamp and Reggiani (1988), Rosser (1991), Varian (1978), Vendrik (1980) and Wilson (1981). A weak element in catastrophe models is the fact that the identification and explanation of turning points is fraught with difficulties, so that the operational basis of catastrophe theory remained feeble. The main problem in analyzing turning points and nonlinear growth patterns is the fact that such phenomena occur in a very irregular fashion, so that normal time series are often inadequate in providing sufficient statistical evidence. Thus instability can **ex post** be traced, but is very hard to predict **ex ante** (Peters, 1988). Prediction is apparently only meaningful in case of stable evolution.

In light of these observations **chaos theory** has in recent years become a popular approach for analyzing unstable phenomena. The next section will offer a sketch of the road towards chaos.

1.3 DYNAMIC SYSTEMS AND CHAOS THEORY
1.3.1 Structural instability

The interest in the dynamics of real-world systems is not recent, but can already be found in early analytical dynamics. Analytical dynamics aims to describe and explain transformations in complicated real-world systems. It emerged about 300 years ago with the concept and calculus of differential equations developed by Newton and Leibniz. Most theories in physics (such as Einstein's relativity theory or Maxwell's electro-magnetic theory) can be represented by means of differential equations.[2] The use of ordinary differential equations in economics has a shorter life span than in physics and dates back to Walras in 1874 and later on Ramsey in the 1920s. However, it was Samuelson (1947) who introduced in economics the new concept of 'dynamic analysis', as a substitute for the commonly accepted concept of 'static analysis', by pointing at Frisch's article "Propagation Problems and Impulse Problems in Dynamic Economics" (1933) (see also Medio, 1979). In this context the notion of stability in economics emerged in the late 1950s (see, for example, Allen, 1959). It should be noted at this stage that stability concepts are in economics strictly linked to dynamic equilibrium phenomena.

Roughly speaking, following Zhang (1990, p. 19) we may define equilibrium "as the state of a dynamic system at which all variables are invariant with respect to time". Then we may assert that "an equilibrium x is stable if all nearby solutions stay nearby. It is asymptotically stable if all nearby solutions not only stay nearby, but also tend to x" (cf. Hirsch and Smale, 1974, p. 180).[3] Inversely "a system is considered to be at non equilibrium if the variables have non-zero rates of change with respect to time. Limit cycles, aperiodic behaviour and chaos (which will be described subsequently) are examples of non-equilibria" (see again Zhang, 1990, p. 19).

In the past decades differential equations have been extensively employed as powerful analytical tools by economic researchers. However, it is often impossible to find the explicit solution of nonlinear differential equations (i.e., equations where the coefficients of the state variables are not constant), unless one resorts to approximations via linear equations. But this means that unexpected discontinuities, a multiplicity of (dis)equilibria as well as a multiplicity of bifurcations and catastrophe behaviour are missed out. More modern approaches try to overcome the above limitation by analyzing not just the properties of particular solutions of differential equations, but **the sensitivity of solutions to perturbations (or their stability)**, in other words, their **structural (in)stability**.[4]

In other words, structural instability means the possible existence of significant qualitative changes in the behaviour of the system (i.e., in the state variables) which are closely connected with bifurcation and catastrophe phenomena that (can) occur when the parameters values (i.e., the control variables) are changing (see for a brief review on bifurcation theory and catastrophe theory applied to spatial-economic patterns among others Day, 1985, Dendrinos and Mullally, 1985, and Zhang, 1991).

It is well known from the literature that a simple bifurcation can transform the equilibrium points from one state (e.g., stable) to another one (e.g., unstable) (or vice versa). An example of such structural instability is provided by a prey-predator system, since a bifurcation (depending on the parameter values) can transform the equilibrium point from a stable 'focus' into an unstable 'focus' via a 'centre' (for a precise description we refer to Nijkamp and Reggiani, 1992a, Ch. 6).

In conclusion, it is possible to analyze, by means of the concept of structural stability, various types of complex dynamics in real world systems, such as the evolution of cities (see Mees, 1975), regional development processes (Anderson and Batten, 1988, and Wilson, 1981), dynamic choices in spatial systems (Fischer et al., 1990) or economic equilibrium analysis (Turner, 1980).

It should be added that a further analysis of dynamics in complex systems should focus on the time-dependent behaviour of the system (compared to the phase portrait or the bifurcation diagrams plotting the state variable values against the parameter values). In this context the identification of (a)periodic motions (or oscillations in the state variables) results as an interesting outcome.

1.3.2 (A)periodic motions in behaviour

Periodic or cyclic solutions are very common in economics (see, for example, the wealth of business cycle literature reviewed by Lorenz, (1989), Kelsey, (1988), Puu, (1989), and Zhang, (1991) as well as in ecology (see, e.g, Haken, 1983a and 1983b), and recently also in geography and regional science (see, e.g., Dendrinos and Mullally, 1985, Nijkamp, 1987, and Nijkamp and Reggiani, 1990).

From a mathematical point of view periodic solutions are strictly connected with the existence of **closed orbits** and **limit cycles.** In the first case states are repeated from one orbit to the next; in the second one this does not hold, but then the orbits are asymptotically close to a closed orbit. In other words, a limit cycle (cf. Lorenz, 1989 for a proper mathematical definition) is a closed orbit (emerging as an equilibrium solution from two-dimensional differential equations) which is also an **attractor** (i.e., a bounded region towards which every trajectory of a dynamic system tends to move). Thus a limit cycle is "closed so that a point moving along the curve will return to its starting position at fixed time intervals and thus execute periodic motion" (cf. Zhang, 1990, p. 207).

It should be noted that there is also the possibility of so-called aperiodic or 'chaotic' oscillations when the amplitude and period vary without any trace of a recognizable pattern (see also next subsections).[5]

A method to find limit cycles in the two-dimensional plane is provided by the Poincaré-Bendixson theorem (see e.g. Haken, 1983a). The Poincaré-Bendixson theorem states in particular the existence of two kinds of attractors for a two-dimensional flow: (1) stable fixed points (see for a classification of fixed points, Nijkamp and Reggiani, 1992a) and (2) periodic solutions or limit cycles. Consequently, the chaotic motion, which is associated with the so-called **strange attractor**, as we will show hereafter, is not possible for two-dimensional flows (see also Lichtenberg and Lieberman, 1983).

Now, it should be noted that the emergence of a limit cycle around a fixed point (if a single parameter varies) is a phenomenon known as **Hopf bifurcation**. It is interesting to note that in an at least three-dimensional system the Hopf bifurcation can be associated with chaotic behaviour (see Marsden and McCracken, 1976). In this framework Newhouse et al. (1978) have shown that after three Hopf bifurcations regular motion becomes highly unstable in favour of motion to a **strange** attractor (defined here as a bounded region of phase space in which initially close trajectories separate exponentially such that the motion becomes chaotic). It is thus clear that central in the analysis of dynamical systems is also the idea of an attractor, i.e. a set of states in the state space of the system which attracts all trajectories emanating from neighbouring points.

1.3.3   (Strange) attractors

Given the above definitions of attractors, it is clear that the attractor does not describe change over time. Indeed, for many parameter values of a dynamical system its attractors (if they exist) are regular objects described by equilibrium points or closed curves. For other parameter values they may have an extremely complex structure. Such attractors are called "strange", a term coined by Ruelle and Takens (1971) in order to indicate an "exponential separation of orbits" (see Day, 1985, and Eckmann and Ruelle, 1985). Strange attractors are strictly connected to chaotic motion (see Annex 6B in Nijkamp and Reggiani, 1992a, for a review of major classes of strange attractors). In particular, in 1975 Li and Yorke christened '**chaotic**' a system with strange attractors or a dynamic situation exhibiting aperiodic - though bounded - trajectories.

A more strict definition of a strange attractor is provided by Schuster (1988, pp. 105-106) as follows:

" a) It is an attractor, i.e., a bounded region of phase space to which all sufficiently close trajectories from the so-called basin of attraction are attracted asymptotically for long enough times. We note that the basin of attraction can have a very complicated structure. Furthermore, the attractor itself should be indecomposable; i.e., the trajectory should visit every point in the attractor in the course of time. A collection of isolated fixed points is no single attractor.

b) The property which makes the attractor strange is the sensitive dependence on the initial conditions; i.e, despite the contraction in volume, lengths need not shrink in all direction, and **points which are arbitrarily close initially, become exponentially separated at the attractor for sufficiently long times.**"

Thus strange attractors (together with the other attracting regions in phase such as fixed points and limit cycles) are pertinent to dissipative[6] systems for which volume elements in phase space shrink with increasing time.

Strange attractors are also connected with the notion of **fractal** dimensions. Fractal is a term coined by Mandelbrot (1977), in order to illustrate that similar structures may repeat themselves at higher orders or dimensions. Essentially, a fractal set is a set having the property of being invariant at different scales (self-similarity and irregularity property) and having a non-integer, fractional dimension (for an application of fractal geometry to urban structure see, e.g., Batty and Longley, 1986, and Frankhauser, 1991). Therefore, the notion of fractals only refers to the geometry of attractors (see also Mandelbrot, 1977, and Peitgen and Richter, 1986).

Since the notion of strange attractors usually refers to the dynamics of the attractors (and not just to their geometry), strange attractors need not have a fractal structure and attractors with a fractal structure need not be chaotic (see, e.g., Holden and Muhamad, 1986).

The concept of 'chaos' results in a straightforward way from the previous basic concepts. Its significance rests essentially on the possibility of studying a variety of natural and social science based models where non-differentiability is typical.

1.3.4 Chaos

During the 1980s chaos theory "received widespread prominence and some scientists have even placed it alongside two other great revolutions of physical theory in the twentieth century - relativity and quantum mechanics. While those theories challenged the Newtonian system of dynamics, chaos has questioned traditional beliefs from within the Newtonian framework" (Crilly, 1991, p. 193).

The interesting characteristic feature of chaos theory is that it addresses essentially the (in)stability of **deterministic, nonlinear dynamic** systems which are able to produce **complex motions** of such nature that they are sometimes seemingly random. In particular such systems incorporate the property that small uncertainties may grow exponentially (although all time paths are bounded), leading to a broad spectrum of different trajectories in the long run, so that precise or plausible predictions are - under certain conditions (see below) - almost impossible. This phenomenon, which is typical of chaotic dynamics, is known as **sensitivity to initial conditions**. Already at the beginning of this century it was recognized by Poincaré that "it may happen that small differences in the initial conditions produce very great ones in the final phenomena. A small error in the former will produce an enormous error in the latter. Prediction becomes impossible, and we have the fortuitous phenomenon" (Poincaré, 1913, p. 397).

Thus, predictability of long term behaviour may become problematic for those nonlinear systems which incorporate chaotic dynamics.

Consequently, the new logic which has emerged in the field of nonlinear dynamics by the introduction of the theory of chaos has also an interesting psychological appeal; model builders need not necessarily be blamed any more for wrong predictions, as errors in predictions may be a result of the system's complexity, as can be demonstrated by examining more carefully the properties of the underlying nonlinear dynamic model structure. However, even though from a behavioural point of view chaos is sometimes identified with 'dynamical stochasticity', 'self-generated noise', or 'intrinsic stochasticity' (see, for example, Hao, 1984), from a mathematical point of view, there is still some uncertainty regarding the precise definition of chaos: "clearly there is still a lot of 'chaos' in chaos theory" (Rosser, 1991, p. 30). On the one hand, chaos is identified by many authors as aperiodic behaviour (see, for example, Guckenheimer, 1979 and Nusse, 1987), starting from Li and Yorke (1975) who, in their well-known theorem 'Period Three Implies Chaos', identify chaos - for any continuous mapping of a one-dimensional interval onto itself - with the existence of cycles of all orders and a scrambled set in which all trajectories are non-periodic and symptotically unstable. In this spirit we recognize also the 'Feigenbaum route' and the 'intermittency route' to chaos. Feigenbaum (1978) in particular discovered the phenomenon of a cascade of bifurcations, each leading to period doubling sequences as a system approaches chaos. Manneville and Pomeau (1979) found that chaos can be intermittent, i.e., chaos can emerge and then be replaced by a new zone of stable equilibrium.

On the other hand, chaos can be identified with the existence of strange attractors, like in the 'Ruelle-Takens-Newhouse route' (Ruelle and Takens, 1971, and Newhouse et al., 1978). These authors showed in particular that after three Hopf bifurcations, it is 'likely' that regular motion becomes highly unstable in favour of motion to a strange attractor.

However, Eckmann and Ruelle (1985) argued that it is the sensitive dependence on initial conditions which is the **true** meaning of chaos, as indicated by the presence of positive Liapunov exponents (measuring the exponential separation of adjacent points). In this context it appears fundamental to have a characterization of chaos by measuring the degree to which a dynamical system is chaotic. The K Kolmogorov entropy (see e.g. Schuster, 1988) is probably the most important approach in this respect, since it is proportional to the rate at which information about the state of the dynamical system is lost in the course of time. In other words, K is the mean rate at which two distinct, but empirically indistinguishable starting points, produce - as time passes - trajectories which are distinguishable. It is clear that K is connected with a positive Liapunov exponent. In particular the latter index quantifies the stretching and contracting in various directions, while the former measures an aggregate of the stretching.

Thus it is a more 'rigorous' and computationally easier approach to test chaos by means of a positive K or a positive Liapunov exponent instead of using the Li-Yorke theorem, the Feigenbaum route or the other routes to chaos.

It should be added, however, that in an empirical context Liapunov exponents do not seem to be sufficient in providing evidence for chaos (see Brock, 1986). In particular

Brock proposed - by pointing at a striking property of chaotic equations (i.e., invariance to linear transformations) - a residual test for economic time-series, based on the following method. If one carries out a linear transformation of chaotic data, then both the original and the transformed data should have the same Liapunov exponent as well as the same correlation dimension[7]. Thus if these indices would appear to be substantially different, then the hypothesis of a deterministic law should be suspicious (see also Frank and Stengos, 1988). This test addresses also the issue of random disturbances in economic activities. There are many independent sources of noise which affect economic data. It is however possible that some of the observed noise is not 'extrinsic noise', independent of the economic system, but 'intrinsic noise' generated by chaotic dynamics of the system itself (see Kelsey, 1988).

It is interesting to note in this context that it has recently been shown that chaotic behaviour is less sensitive to noise than periodic orbits (see again Kelsey, 1988). Because of this low sensitivity of chaotic equations to random errors, it turns out to be difficult to identify the nature of random disturbances in time series.

In conclusion, chaos should not be regarded any more as a peculiar theoretical possibility, but as a generic property of the solution set of nonlinear dynamic equations which is suitable for empirical research. Examples of chaotic empirical phenomena can be found, inter alia, in solar pulsation (Kurths and Herzal, 1986), cardiac cells (Glass et al., 1983), long-term climatic change (Nicolis, 1986), measles epidemics (Schaffer and Kot, 1985), hydrodynamic turbulence (Swinney, 1983), and biological and physiological systems such as nephrons, neural and metabolitic-networks (Degn et al., 1987). The interest in economics and regional science is also rapidly growing. In conclusion, chaos "has proved to provide a fruitful approach to organizing one's thoughts concerning many observed phenomena" (Frank and Stengos, 1988, p. 104). For further expositions on chaos theory we refer to Chapter 2 of the present volume.

1.4 EVOLUTION AND ECOLOGICALLY-BASED SYSTEMS: A NICHE APPROACH

1.4.1 Evolution as complex hierarchical organization

One of the recent outstanding challenges to social science research is the burgeoning dialogue between economics (and other social sciences on the one hand) and ecology/biology on the other hand. Firstly, the persistent dialectic in biology between evolution conceived of as the result of historical processes and natural selection (following Darwin, 1859) and evolution conceived as an 'intelligible' structure (see Russell, 1916 and Waddington, 1957) attempting "to describe necessary and sufficient conditions for the production of a particular state or structure, whereas historical explanations are necessarily restricted to description of some necessary conditions only" (cf. Goodwin, 1992, p. 90) is clearly reflected in recent discussions in economic and social sciences. A synthesis between the two approaches would require - from a formal viewpoint - that the **dynamics of development be adequately described,** followed by a proper analysis of the **dynamic stability of the generated states** (a situation to which the concept of 'natural selection' refers; see again Goodwin, 1992). Such conditions will undoubtedly also be relevant for the formal apparatus of nonlinear dynamic economics.

Secondly, it is noteworthy that the ongoing debate in ecology and biology between **gradualism** (continuous evolution) and **saltationism** (discontinuous evolution) has also been joined by various economists, for instance, respectively by Marshall (1920) in his study of the evolution of market structure, and by Schumpeter (1934) in his analysis of innovative activities of entrepreneurs. Positioned in between these two main streams of evolutionary theories is Boulding (1978), who argued for a mix of continuous and discontinuous processes in which instability can play a significant role by generating discontinuous events.

In understanding the nature of the evolution of spatial economic systems the theory of chaos has brought to light two important results:

i) (more or less) nonlinear dynamical systems possess parameter regimes within which dynamics is essentially random;

ii) although, by definition, seemingly chaotic systems do not display any systematic features, that is actually not the case. There is 'order in chaos', witness the universal number by Feigenbaum or the fractal geometry of strange attractors.

Thus it seems plausible that **constraints** in models leading to particular parameter values or to different connections (organizations) in individual components of a system play a crucial role (see Davies, 1992). In other words, chaos models provide a new challenge to the **reductionist** view that the global behaviour of a system can in general be deducted from knowledge on individual components. On the other hand, the relevance of **organizational aspects** of a dynamic system suggests the key element in the **holistic** view, that at each level of complexity in the hierarchy of systems new qualities emerge that are not only absent, but clearly meaningless at lower levels (Davies, 1992). As Davies (1992, p.107) points out: "These **constrained system models of complexity** are the **modern frontier** of theoretical model-building, imagining the real (conceded to be complex) world in a new type of mathematics, in which concepts such as non time-reversible dynamics, algorithmic complexity and global patterning make the running. The old type of model-building dwelt on simplicity and relied heavily on reduction, and imaged the world (in an idealised, simplified way) by continuous curves and processes, reversible dynamics and the language of calculus".

Starting from the above recent issues on the principles of evolution we underline here the essential role of **hierarchical organization** (in the evolution of a spatial-economic system) leading to a self-organizing process as the result of competition, substitution or complementarity among different subsystems within the main system.

It is interesting to note that such a structure can be modelled by means of a recent and increasingly popular approach to competitive analysis (and network analysis in general) in ecology and social sciences, viz. **niche theory**.

1.4.2 Evolution of hierarchical systems by means of niche theory

It has been recently shown (see Rosser, 1991) how the analysis of complex (i.e. non-smooth) phenomena in economics may strongly rely on ecologically-based models, in particular the May (1976) equation, exhibiting bifurcations and chaos for one-dimensional systems and the Lotka-Volterra models (Lotka, 1920 and Volterra, 1931) which are able to display (non)cyclical oscillations for n-dimensional systems.

While being capable of integrating these two families of models, niche theory has recently been advocated by economists and geographers (see Nijkamp and Reggiani, 1992b, 1992c) as a relevant tool for modelling hierarchies of levels of self-organization. Briefly, niche theory can be formulated as follows:

$$\dot{x}_i = x_i (k_i - \sum_{j=1}^{n} \alpha_{ij} x_j) \tag{1.1}$$

where x_i represents the population of a species i (i=1, 2,..n), the constant k_i is a function of the carrying capacity for the ith species, and α_{ij} represent the competition coefficients between species, measuring the amount of resource sharing or **niche overlap**. In other words, the niche concept expressed in (1.1) refers to the optimal adjustment process in dynamic systems with scarce resources (usually formalized by means of bell-shaped utilization functions; see May, 1973) where the value of niche overlapping generates the possibility of extinction or coexistence of species (see, for details, Nijkamp and Reggiani, 1992b). Thus, equations based on formulation (1.1) and applied to economic-spatial systems can be interpreted by means of niche theory. For example, niches occupied successively by species of increasing effectiveness form the basis of models by Allen and Sanglier (1981) and Camagni et al. (1986), explaining the appearance of new economic functions in urban dynamics, by Johansson and Nijkamp (1987) studying urban and regional development with competing regions or by Grübler and Nakicénovic (1991) describing the substitution of transport infrastructure by means of empirical diagrams.

It is interesting to note that the hierarchical levels of species emerging from niche theory (where each of them can be connected with further chains of hierarchical niches) leads to the concept of evolution as described by Prigogine and Stengers (1984, pp. 193-194): "Living societies continually introduce new ways of exploiting resources or of discovering new ones (that is k_i increases) and continually discover new ways of extending their lives or multiplying more quickly. Each ecological equilibrium defined by the logistic equation is thus only temporary, and a logistically defined niche will be occupied successively by a series of species, each capable of ousting the preceding one when 'aptitude' for exploiting the niche becomes greater".

Formally this 'aptitude' is determined by the value of x_i in expression (1.1) obtained from the condition $\dot{x}_i = o$. Empirically, this 'aptitude' reveals the fundamental feature for a new species (e.g., a new technological innovation in the economic market) replacing the old one by being not only able to do perform the same functions, but to generate also new opportunities (see Jantsch, 1980).

Generally niche evolution can also be chaotic[8]. The possibility of chaotic behaviour in a niche chain has been recently pointed out by Nijkamp and Reggiani (1992b, 1992c) - with reference to a transport system - by exploring the impact of an unstable evolution of an irregular 'niche' on the system as a whole.

This brings us back to the present volume, since the theme that unites the contributions to this volume is essentially this problem of understanding the relationships between order and complexity in the evolution of socioeconomic systems. Thus the various publications in this book define a research programme involving a search for the principles of evolutionary theories underlying spatial-economic processes as well as their relationships to their fundamental requirements in specific space-time environments.

ACKNOWLEDGEMENT

The present paper has been developed under funding of the Italian C.N.R. grant no. 92.03435.CT11.

REFERENCES

Allen, R.G.D., 1959, **Mathematical Economics**, MacMillan, London (2nd edition).

Allen, P. and M. Sanglier, 1981, Urban Evolution, Self-Organization, and Decision-Making, **Environment and Planning A**, vol. 13, pp. 167-183.

Andersson, A.E. and D.F. Batten, 1988, Creative Nodes, Logistical Networks and the Future of the Metropolis, **Transportation**, vol. 14, pp. 281-293.

Batten, D.J., J. Casti and B. Johannson (eds.), 1987, **Economic Evolution and Structural Adjustment**, Springer-Verlag, Berlin.

Batty, M. and P.A. Longley, 1986, The Fractal Simulation of Urban Structure, **Environment and Planning A**, vol. 18, pp. 1143-1179.

Boulding, K.E., 1978, **Ecodynamics: A New Theory of Social Evolution**, Sage, Beverly Hills.

Brock, W.A., 1986, Distinguishing Random and Deterministic Systems: Abridged Version, **Journal of Economic Theory**, vol. 40, pp. 168-195.

Camagni, R., L. Diappi and G. Leonardi, 1986, Urban Growth and Decline in a Hierarchical System, **Regional Science and Urban Economics**, vol. 15, pp. 145-160.

Casti, J., 1979, **Connectivity, Complexity, and Catastrophe in Large-Scale Systems**, J. Wiley, New York.

Crilly, T., 1991, The Roots of Chaos. A Brief Guide. In Crilly, A.J., R.A. Earnshaw and H. Jones (eds.), **Fractals and Chaos**, Springer-Verlag, Berlin.

Darwin, C., 1859, **The Origin of Species**, Penguin, Harmondsworth.

Davies, P.C.W., 1992, The Physics of Complex Organization, **Theoretical Biology. Epigenetic and Evolutionary Order from Complex Systems** (Goodwin B. and P. Saunders, eds.), John Hopkins Press Ltd., London, pp. 101-111.

Day, R.H., 1985, Dynamical Systems Theory and Complicated Economic Behaviour, **Environment and Planning B**, vol. 2, pp. 55-64.

Day, R.H. and J.-L. Walter, 1990, Economic Growth in the Very Long Run, in: Barnett, W.A., J. Geveke and K. Shell (eds.), **Economic Complexity; Chaos, Sunspots, Bubbles and Nonlinearity**, Cambridge University Press, Cambridge, pp. 253-289.

Degn, H., A.V. Holden and L.F. Olsen (eds.), 1987, **Chaos in Biological Systems**, Plenum Press, New York.

Dendrinos, D.S. and H. Mullally, 1985, **Urban Evolution. Studies in the Mathematical Ecology of Cities**, Oxford University Press, Oxford.

Dendrinos, D.S. and J.B. Rosser, 1992, Fundamental Issues in Non Linear Urban Population Dynamic Models, **The Annals of Regional Science**, vol. 26, no. 2, pp. 111-134.

De Palma, A. and C. Lefèvre, 1987, The Theory of Deterministic and Stochastic Compartmental Models and its Applications, in C.S. Bertuglia, G. Leonardi, S. Occelli, G.A. Rabino, R. Tadei and A.G. Wilson (eds.), **Urban Systems: Contemporary Approaches to Modeling**, Croom Helm, London, pp. 490-540.

Eckmann, J.P. and D. Ruelle, 1985, Ergodic Theory of Chaos and Strange Attractors,

Review of Modern Physics, vol. 57, no. 3, pp. 617-656.

Feigenbaum, H.J., 1978, Quantitative Universitality for a Class of Non-Linear Transformations, **Journal of Statistical Planning**, vol. 19, pp. 25-52.

Fischer, E.O. and W. Jammernegg, 1986, Empirical Investigation of a Catastrophe Exclusion of the Phillips Curve, **Review of Economics and Statistics**, vol. 68, no. 1, pp. 9-17.

Fischer, M.M., P. Nijkamp and Y.Y. Papegeorgiou, 1990, **Spatial Choices and Processes**, North-Holland, Amsterdam.

Forrester, J., 1968, **Principles of Systems**, Wright-Allen Press, Cambridge, Mass.

Frank, M. and T. Stengos, 1988, Chaotic Dynamics in Economic Time-Series, **Journal of Economic Surveys**, vol. 2, no. 2, pp. 103-133.

Frankhauser, P., 1991, Aspects Fractals des Structures Urbaines, **L'Espace Géographique**, no. 1, pp. 45-69.

Frisch, R., 1933, Propagations Problems and Impulse Problems in Dynamic Economics, **Economic Essays in Honour of Gustav Cassel**, Allen & Unwin, London.

Glass, L., M.R. Guevera, M.R. Shrier and R. Perez, 1983, Bifurcation and Chaos in a Periodically Stimulated Cardiac Oscillator, **Physica D**, vol. 7, pp. 89-101.

Goodwin, B.C., 1992, Evolution and the Generative Order, **Theoretical Biology. Epigenetic and Evolutionary Order from Complex Systems** (Goodwin B. and P. Saunders, eds.), John Hopkins Press Ltd., London, pp. 89-100.

Grassberger, P. and I. Procaccia, 1983a, Measuring the Strangeness of Strange Attractors, **Physica D**, vol. 9, pp. 189-208.

Grassberger, P. and I. Procaccia, 1983b, Estimation of the Kolmogorov Entropy from a Chaotic Signal, **Physical Review A**, vol. 28, pp. 2591-2593.

Grübler, A. and Nakicenovic, 1991, **Evolution of Transport Systems: Past and Future**, RR-91-8, International Institute For Applied Systems Analysis, Laxenburg.

Guckenheimer, J., 1979, Sensitive Dependence to Initial Conditions for One-Dimensional Maps, **Communications in Mathematical Physics**, vol. 70, 1979, pp. 133-160.

Haken, H., 1983a, **Synergetics**, Springer-Verlag, Berlin.

Haken, H., 1983b, **Advanced Synergetics**, Springer-Verlag, Berlin.

Hao, B.-L. (ed.), 1984, **Chaos**, Scientific Publication. Co., Singapore.

Hirsch, M. and S. Smale, 1974, **Differential Equations, Dynamical Systems and Linear Algebra**, Academic Press, London.

Holden, A.V. and M.A. Muhamad, 1986, A Graphical Zoo of Strange and Peculiar Attractors, in: A.V. Holden (ed.), **Chaos**, Manchester University Press, Manchester, pp. 15-35.

Jantsch, E., 1976, Self-Transcendence: New Light on the Evolution Paradigm, **Evolution and Consciousness. Human Systems in Transition** (Jantsch E. and C.H. Waddington, eds.), Addison Wesley, Reading, Massachussetts, pp. 9-10.

Jantsch, E., 1980, **The Self-Organizing Universe**, Pergamon Press, Oxford.

Johansson, B. and P. Nijkamp, 1987, Analysis of Episodes in Urban Event Histories, **Spatial Cycles** (Van den Berg, L., L.S. Burns and W. H. Klaassen, eds.), Gower, Aldershot, pp. 43-66.

Kelsey, D., 1988, The Economics of Chaos or the Chaos of the Economics, **Oxford University Papers**, vol. 40, pp. 1-3.

Kurths, J. and H. Herzel, 1987, An Attractor in a Solar Time Series, **Physica D**, vol. 25, pp. 165-172.

Li, T.Y. and J.A. Yorke, 1975, Period Three Implies Chaos, **American Mathematical Monthly**, vol. 82, no. 10, pp. 985-992.

Lichtenberg, A.J. and M.A. Lieberman, 1983, **Regular and Stochastic Motion**, Springer-Verlag, Berlin.

Lombardo, S.T. and G.A. Rabino, 1983, Some Simulations of a Central Place Theory Model, **Sistemi Urbani**, vol. 5, pp. 315-332.

Lorenz, H.-W., 1989, **Non-Linear Dynamical Economics and Chaotic Motion**, Lecture Notes in Economics and Mathematical Systems, vol. 334, Springer-Verlag, Berlin.

Lotka, A., 1920, Analytical Notes on Certain Rhythmic Relation in Organic Systems, **Proceedings of the National Academy of Sciences, United States**, vol. 6, pp. 410-415.

Mandelbrot, B., 1977, **The Fractal Geometry of Nature**, V.H. Dreeman and Company, New York.

Manneville, P. and Y. Pomeau, 1979, Intermittency and the Lorenz Model, **Physics Letters A**, vol. 75, no. 1/2, pp. 1-2.

Marsden, J.E. and M. McCracken, 1976, **The Hopf Bifurcation and its Applications**, Springer-Verlag, Berlin.

Marshall, A., 1920, **Principles of Economics**, Macmillan, London.

May, R.M., 1973, **Stability and Complexity in Model Ecosystems**, Princeton University Press, Princeton.

May, R., 1976, Simple Mathematical Models with Very Complicated Dynamics, **Nature**, vol. 261, pp. 459-467.

Meadows, D.L. and D.H. Meadows (eds.), 1973, **Toward Global Equilibrium: Collected Papers**, Wright-Allen Press, Cambridge, Mass.

Medio, A., 1979, **Teoria Non Lineare del Ciclo Economico**, Il Mulino, Bologna.

Mees, A., 1975, The Revival of Cities in Medieval Europe: An Application of Catastrophe Theory, **Regional Science and Urban Economics**, vol. 5, pp. 403-425.

Newhouse, S., D. Ruelle and F. Takens, 1978, Occurrence of Strange Axiom-A Attractors near Quasiperiodic Flow or T^m, $m \geq 3$, **Communications in Mathematics Physics**, vol. 64, pp. 35-40.

Nicolis, G., 1986, Dissipative Systems, **Reports on Progress in Physics**, vol. 49, pp. 873-949.

Nicolis, G., and I. Prigogine, 1977, **Self-Organization in Nonequilibrium Systems**, Wiley, New York.

Nijkamp, P., 1987, Long Term Economic Fluctuations: A Spatial View, **Socio-Economic Planning Sciences**, vol. 21, no. 3, 189-197.

Nijkamp, P. and A. Reggiani, 1990, An Evolutionary Approach to the Analysis of Dynamic Systems with Special Reference to Spatial Interaction Models, **Sistemi Urbani**, vol. 1, pp. 601-614.

Nijkamp, P. and A. Reggiani, 1992a, **Interaction, Evolution and Chaos in Space**, Springer Verlag, Berlin.

Nijkamp, P. and A. Reggiani, 1992b, Spatial Competition and Ecologically Based Socio-Economic Models, **Socio Spatial Dynamics**, vol. 3, No. 2, pp. 89-109.

Nijkamp. P. and A. Reggiani, 1992c, Impacts of Changing Environmental Conditions on Transport Systems, Research Memorandum 1992-51, Dept. of Economics, Free University, Amsterdam.

Nusse, H.E., 1987, Asymptotically Periodic Behaviour in the Dynamics of Chaotic Mapping, **SIAM Journal of Applied Mathematics**, vol. 47, pp. 498-515.

Peitgen, H.O. and P.H. Richter, 1986, **The Beauty of Fractals**, Springer-Verlag, Berlin.

Peters, T., 1988, **Thriving on Chaos**, MacMillan, London.

Poincaré, H., 1913, **The Foundations of Science. Science and Method**, (English Translation: The Science Press, Lancaster, 1946).

Poston, T. and I. Stewart, 1978, **Catastrophe Theory and its Applications**, Pitman, London.

Power and Air, **Biological Bulletin**, vol. 25, pp. 79-120.

Prigogine, I. and I. Stengers, 1984, **Order out of Chaos**, Fontana, London.

Puu, T., 1989, **Non-Linear Economic Dynamics**, Springer-Verlag, Berlin.

Rosser, J.B., 1991, **From Catastrophe to Chaos: A General Theory of Economic Discontinuities**, Kluwer Academic Publishers, Dordrecht.

Ruelle, D. and F. Takens, 1971, On the Nature of Turbulence, **Communications in Mathematical Physics**, vol. 20, pp. 167-192.

Russel, E.S., 1915, **Form and Function**, Murray, London.

Samuelson, P.A., 1947, **Foundations of Economic Analysis**, Harvard University Press, Cambridge, Mass.

Schaffer, W.M. and M. Kot, 1985, Nearly One-Dimensional Dynamics in an Epidemic, **Journal of Theoretical Biology**, vol. 112, pp. 403-427.

Schumpeter, J.A., 1934, **The Theory of Economic Development**, Harvard University Press, Cambridge, Mass.

Schuster, H.G., 1988, **Deterministic Chaos**, VCH, Veinheim.

Swinney, H., 1983, Observations of Order and Chaos in Non-Linear Systems, **Physica D**, vol. 7, pp. 3-15.

Thom, R., 1972, **Structural Stability and Morphogenesis**, Addison-Wesley, Reading.

Varian, H., 1978, Catastrophe Theory and the Business Cycle, **Economic Inquiry**, vol. 17, pp. 14-28.

Vendrik, M.C.M., 1990, Habits, Histeresis and Catastrophes in Labor Supply, Research Memorandum 90-031, Department of Economics, University of Limburg, Maastricht.

Volterra, V., 1931, **Leçons sur la théorie mathématique de la lutte pour la vie**, Gauthier-Villars, Paris.

Waddington, C.H., 1957, **The Strategy of the Genes**, Allen and Unwin.

Walras, L., 1874, **Eléments D'Economies Pure**, L. Corbaz, Lausanne (English Translation: W. Jaffe, Elements of Pure Economics, Allen and Unwin, London, 1954).

Wilson, A.G., 1981, **Catastrophe Theory and Bifurcation**, Croom Helm, London.

Wright, S., 1931, Evolution in Mendelian Population, **Genetics**, vol. 16, pp. 97-189.

Zeeman, F.C., 1977, **Catastrophe Theory: Selected Papers 1972-1977**, Addison-Wesley, Reading.

Zhang, W.B., 1990, **Economic Dynamics**, Springer-Verlag, Berlin.

Zhang, W.B., 1991, **Synergetic Economics**, Springer-Verlag, Berlin.

NOTES

1. Catastrophe is a term coined by Thom (1972) for explaining biological morphogenesis. Thom also gave a topological classification of the elementary types of catastrophes depending on the number of control variables in a dynamical system.

2. Differential equations can in general be formalized as follows:

$$\dot{x}_i = f_i(x_i, x_2, \ldots, x_n; \alpha_i, \alpha_2, \ldots, \alpha_m) \quad i = 1, 2, \ldots, n$$

where the symbol · indicates dx/dt, (i.e. the derivative of function x with respect to the variable t with t being time), and where the α's are relevant parameters.

3. For the classification of different kinds of (in)stability related to a general system of two differential equations, see also Annex 6A in Nijkamp and Reggiani (1992a).

4. For a broad discussion on structural stability in mathematics as well as in topological theory see Thom (1972).

5. An interesting review on illustrations of (a) periodic motions with a special focus on technological change, business cycles and economic growth, and competition can be found in Batten, Casti and Johansson (1987).

6. However one can find 'chaotic' (in the sense of irregular) regions also in conservative systems, but they are not attractors.

7. The concept of correlation dimension was introduced by Grassberger and Procaccia (1983a,b). The correlation dimension measures the ratio between the spatial correlation (between all points on the attractor - for a given radius r) and r (see for details Lorenz,

1989).

8. Formally, this can happen when expression (1.1) is formulated in at least three equations, or if (1.1) is - approximated for a discrete system - in $n \geq 1$ equations.

PART A: NONLINEAR DYNAMICS IN ECONOMICS

CHAPTER 2
LESSONS FROM NONLINEAR DYNAMIC ECONOMICS
Peter Nijkamp and Jacques Poot

2.1 INTRODUCTION

Our economic world is highly dynamic and exhibits a wide variety of fluctuating patterns. This forms a sharp contrast with our current economic toolbox, which is largely filled with linear and comparative static instruments. Clearly, linear economic models do not necessarily generate stable solutions, but their evolution is only capable of generating four types of time paths: oscillatory and stable; oscillatory and explosive; monotonic and stable; and monotonic and explosive. This is true for linear models of any order, so that such models are only able to generate a limited spectrum of dynamic behaviour. Non-periodic evolution for instance, can normally not be described by our analytical apparatus, unless stochastic processes describing nonlinear transition processes are assumed (see Brock, 1986, Priestley, 1988, and Schuster, 1984).

Nonlinear dynamic relationships in economics are certainly not an unknown phenomenon and Goodwin's business cycle model of the 1950s is a well known example (see also Goodwin, 1982), but in most empirical applications linear (or linearized) models are still dominant. One important reason is that **nonlinear dynamic econometrics** is by no means a well developed field of research and another is that **specification theory** is still a weak part in economic modelling (see Blommestein, 1986). In general, the issue of nonlinear dynamics in economic modelling is less interesting when it concerns stochastic properties of the system, but much more when it concerns the way synchronic and diachronic processes are intertwined (see also Barnett et al., 1990, Lichtenberg and Lieberman, 1983, Liossatos, 1980, and Turner, 1980). Discontinuities in a system's behaviour may then emerge under certain conditions, which reflects essentially a morphogenesis in the evolution of the system concerned. Such morphogenesis may be based on either endogenous forces (e.g., behavioural feedbacks, overlapping generations), or exogenous forces (e.g., in the case of random shock models or ceilings and floors models) or a combination of both (e.g., regime switching models).

In recent years, a wide variety of dynamic economic models for countries, sectors or regions has been developed. Surprisingly enough, only a limited number of these studies exhibited **structural** dynamics. A major analytical problem in this respect is the question whether structural changes are caused by **intra-systemic** (endogenous) developments or exogenous forces (external to the system). This problem bears some similarities to the well known scientific debate on the existence of long waves in economics, where especially the Schumpeterian viewpoint regarding the **endogeneity** of phases in a Kondratieff long wave is being tested (see also Grandmont, 1985, Kleinknecht, 1986).

In any case, a meaningful model for analyzing and predicting structural dynamics of an economic system should be able to generate various trajectories for the evolution of the system, in which both endogenous and exogenous fluctuating patterns may play a role. Furthermore, such a model may lead to testable hypotheses in order to explore under which conditions a stable development may emerge. In recent years, this has led to the popularity of the **theory of chaos**, especially since this new research line is focusing attention on the driving forces and trajectories of dynamic evolution.

In this paper we address situations in which nonlinear dynamics may arise in economic phenomena and we will focus on the predictive ability of models which exhibit nonlinear motion. In the next section we review key issues in the theory of chaos. Since a number of comprehensive surveys of the economic applications of this theory in the last decade have been published recently (see Kelsey, 1988; Baumol and Benhabib, 1989; Boldrin and Woodford, 1991; Scheinkman, 1990; Radzicki, 1990; Rosser, 1991), our survey can be brief. However, we will illustrate the key issues by means of two models of economic development in discrete time. Section 2.3 describes a simple nonlinear growth model in which growth is generated by exogenous accumulation of conventional production factors and social overhead capital. In this model, bottlenecks resulting from congestion and other externalities, generate decreasing returns to the conventional production factors. For plausible parameter values, this process leads to monotonic convergence to a stationary state. However, under the assumption that productivity shocks are discrete and lumpy, cyclic or chaotic motion may emerge.

In the second model, described in section 2.4, endogenous technological change provides a positive feedback to economic growth, but bottleneck phenomena again limit growth. Stylized facts regarding economic development give here little guidance about certain parameter values but it will be shown that the model could exhibit a wide range of dynamic behaviour.

In the last section we reflect on the implications for further work in this area.

2.2 ISSUES IN CHAOS THEORY

Chaos theory has attracted widespread interest in the social sciences. In addition to the economic surveys mentioned earlier, informative reviews of chaos theory and its relevance for the social sciences can be found in among others, Andersen (1988), Benhabib and Day (1981, 1982), Boldrin (1988), Crilly et al., (1990), Devaney (1986), Guckenheimer and Holmes (1983), Lasota and Mackey (1985), Lung (1988), Pohjola (1981), Prigogine and Stengers (1985), Rosser (1991), Stewart (1989), and Stutzer (1980).

It is noteworthy that the new logic which has emerged in the area of nonlinear dynamics by the introduction of the theory of chaos has also an interesting psychological appeal; model builders need not necessarily be blamed any more for wrong predictions, as errors in predictions may be a result of the system's complexity, as can be demonstrated by examining more carefully the properties of the underlying nonlinear dynamic model. A fact is that chaos theory is currently regarded as a major discovery with a high significance for both the natural and social sciences.

An important feature of chaos theory is that it is essentially concerned with **deterministic, nonlinear dynamic** systems which are able to produce **complex motions** of such a nature that they are sometimes seemingly random. In particular, they incorporate the feature that small uncertainties may grow exponentially (although all time paths are bounded), leading to a broad spectrum of different trajectories in the long run, so that precise or plausible predictions are - under certain conditions - very unlikely.

In this context, a very important characteristic of nonlinear models which can generate chaotic evolutions is that such models exhibit strong sensitivity to initial conditions. Points which are initially close will on average diverge exponentially over time, although their time path is bounded and they may be from time to time briefly very close to each other. Hence, even if we knew the underlying structure exactly, our evaluation of the current state of the system is subject to measurement error and, hence it is impossible to predict with confidence beyond the very short run. Similarly, if we knew the current state with perfect precision, but the underlying structure only approximately, the future evolution of the system would also be unpredictable. The equivalence of the two situations has been demonstrated by e.g. Crutchfield et al. (1982).

An example of the extreme sensitivity of chaotic models to parameter values is demonstrated in Figure 1. This figure shows an example of the standard May (1976) model $X_{t+1} = a (1-X_t)X_t$ with the parameter $a=3.8$ and the initial value X_0 equal to the equilibrium point $1-1/a$ with as much precision as a modern high-speed computer allows. Figure 1 shows that under these circumstances the model exhibits stability for up to 50 periods, but the finite precision arithmetic of the computer generates after this point slight movements in X_t which are quickly amplified to chaotic fluctuations.

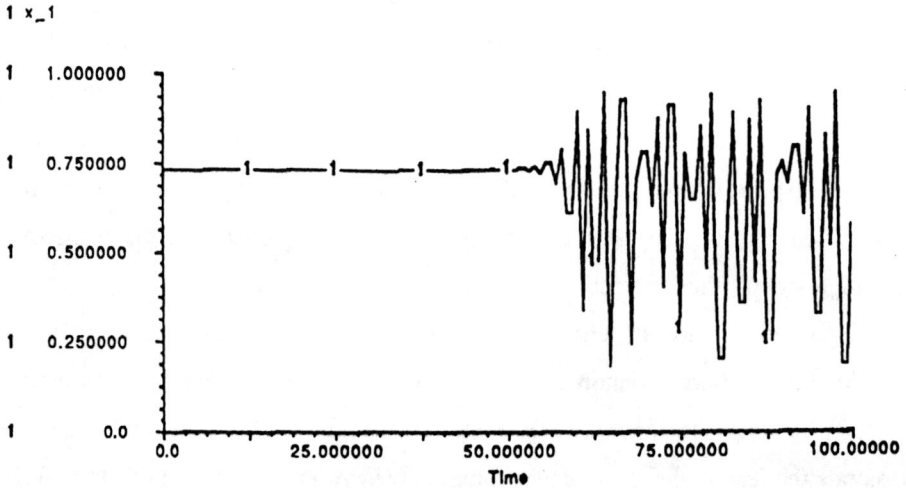

Figure 1. The standard May model.

After a series of interesting studies on chaotic features of complex systems in physics, chemistry, biology, meteorology and ecology, chaos theory has also been introduced and investigated in the field of economics and geography. The main purpose of the use of this theory in the social sciences was to obtain better insight into the underlying causes of unforeseeable evolutions of complex dynamic systems.

In recent years, the economics discipline has witnessed an increasing wave of contributions in the use of chaos theory for analyzing economic dynamics. We noted already in the introduction that in the 1950s an interesting application of chaos theory to economics (in particular, the existence of stable limit cycles in nonlinear dynamics) was developed by Goodwin. He studied economic dynamics by means of an **accelerator-multiplier** framework for persistent, deterministic oscillations as an **endogenous** result of a dynamic economic system (see for a survey Goodwin, 1982). But only recently the awareness has grown that deterministic (periodic or a-periodic) fluctuations (or even bifurcations and jumps) in a complex dynamic economic system may be the result of small perturbations. Unexpected behaviour of nonlinear dynamic models leads to the question of **validity of model specifications** (i.e., are model specifications compatible with plausible economic hypotheses) and of **testability of model results** (i.e., are model results - qualitatively or quantitatively - justifiable from possible nonlinear patterns in the underlying data set). Further expositions on these issues can be found, amongst others, in Scheinkman (1990), Baumol and Benhabib (1989), Baumol and Quandt (1985), Chen (1988), Kelsey (1988) and Lorenz (1989). Applications and illustrations of chaos theory in economics can be found inter alia in the following fields:

- growth and business cycle theory (Balducci et al., 1984; Benhabib and Day, 1982; Boldrin, 1988; Day, 1982; Funke, 1987; Guckenheimer et al., 1977; Grandmont, 1985, 1986; Hommes et al., 1990; Stutzer, 1980)
- cobweb models (Chiarella, 1988)
- long waves analysis (Nijkamp, 1987; Rasmussen et al., 1985; Sterman, 1985)
- R&D analysis (Baumol and Wolff, 1983; Nijkamp et al., 1991)
- consumer behaviour (Benhabib and Day, 1981)
- duopoly theory (Rand, 1978; Dana and Montrucchio, 1986)
- economic competition (Deneckere and Pelikan, 1986; Ricci, 1985)

- international trade (Lorenz, 1987)
- competitive interactions between individual firms (Albin, 1987)
- equilibrium theory (Hommes and Nusse, 1989; Nusse and Hommes, 1990).
- stock returns and exchange rates (LeBaron, 1989).

Interesting applications of chaos theory to related branches of economics can be found in geography and regional science (see for a survey Nijkamp and Reggiani, 1989, 1992). Examples here are:
- regional industrial evolution (White, 1985)
- urban macro dynamics (Dendrinos, 1984)
- spatial employment growth (Dendrinos, 1986)
- relative population dynamics (Dendrinos and Sonis, 1987)
- spatial competition and innovation diffusion (Sonis, 1986, 1987)
- migration systems (Haag and Weidlich, 1983, Reiner et al., 1986)
- urban evolution (Batty 1991, Nijkamp and Reggiani, 1988)
- transport systems (Reggiani, 1990)

It is interesting to observe that most applications of chaos theory in economics (and in general the social sciences) lack empirical content. While empirical research on chaos went hand in hand with theoretical developments in the natural sciences in the early 1980s, attempts to detect chaos in financial and economic data are more recent. The results in this area are so far disappointing. Brock (1989) claims that as yet no class of structural economic models has been **estimated** which allows for chaotic behaviour **and** in which the estimated model parameters are indeed in the chaotic range. Moreover, statistical tests which have been designed to detect chaos in time series without **a priori** specification of the nature of the data generating process, have not provided as yet unambiguous empirical support for the presence of chaos in observable economic processes.

The central concept for the statistical detection of chaos is that of **dimension** of the time series which can be loosely interpreted as the minimum number of lags that one would need to describe the dynamical behaviour of a time series in the long-run (see also

Brock, 1989). A very long truly random time series has a near-infinite dimension, but sequences of observations in a chaotic time series clump together in a lower dimensional space. Several statistical tests have been developed to test for the presence of such low-dimensional deterministic chaos, notably the Brock-Dechert-Scheinkman (1987) statistic, which has a standard normal distribution under the null hypothesis of pure randomness. This statistic detects a wide range of deviations from white noise in fluctuations rather than just chaos and is therefore a useful tool in specification analysis for estimation of time series models. Additional tools such as the largest Lyapunov exponent and recurrence plots are available (for details see Brock, 1988). However, so far the weight of evidence in ranges of economic and financial data is against the hypothesis that there is low-dimensional deterministic chaos in such time series. However, this does not imply that a nonlinear dynamic structure is absent from economic and financial time series, but the available tests are not able to identify the nature of this structure. For example, Brock and Sayers (1988) found evidence of nonlinearity in the following US statistics: employment and unemployment (quarterly), industrial production (monthly) and pigiron production (annually). Empirical evidence of nonlinear dynamics is now also emerging elsewhere, e.g. in weekly price observations in German agricultural markets (Finkenstädt and Kuhbier, 1990) and Austrian demographic data (Prskawetz, 1990). However, nonlinear determinism tends to be absent in many macroeconomic aggregates such as GNP and private investment. One rare finding of chaos in monetary aggregates was recorded by Barnett and Chen (1988), but their finding has been convincingly challenged by Ramsey et al (1990).

There are at least four reasons why linear modelling of economic time series may be adequate and why a nonlinear deterministic structure may therefore be absent or undetectable in such cases. First, farsighted economic agents have a desire to smooth consumption and production over time. This creates **negative feedback** loops. In contrast, it can be easily demonstrated that deterministic chaos requires **positive feedback** loops which may be found in phenomena such as industrial clustering, networking and the growth of cities, but which could be less likely in financial and economic time series (Brock, 1989). For example. the efficient markets hypothesis suggests that if deterministic structure could be detected in **innovations** in e.g. share market data, such structure would vanish as agents would attempt to profitably exploit it in their forecasting of price movements (e.g. Fama, 1976). Some evidence of low-dimensional chaos has nonetheless been found in time series on US stock returns (Scheinkman and LeBaron, 1989), although an analysis of e.g. New Zealand share market data suggested the opposite (Allen, 1989). A third reason for the difficulty in detecting chaos is that economic time series are, after detrending, inherently noisy due to measurement errors and outside shocks. In this case there may be some deterministic structure underlying the stationary fluctuations, but the high-dimensional chaos generated by this process may be indistinguishable from true randomness. Finally, the disappointing results to date may be explained by the focus on relatively short traditional macroeconomic series rather than long time series of microlevel data in which there may be more potential for nonlinear determinism. The power of the available tests may only be satisfactory when the available number of observations is of the order of 10 to 10 rather than 10^2 as is common practice, although it has to be added that recent statistical methods of dimension calculus and nonlinearity testing can get by with much smaller data sets (using mainly Monte Carlo tests; see Hsieh, 1989 and Ramsey et al., 1990). In general however, suitable microlevel data sets are yet hardly available.

In the next two sections we will evaluate the use and relevance of chaos theory by means of two related examples, one in the field of economic restructuring (Section 2.3) and another one concerning the impact of innovation and R&D on diseconomies of scale (Section 2.4). In both cases it will be shown that in case of reasonable growth rates stable behaviour is likely to emerge, but that in case of (very) high growth suddenly unexpected fluctuations may emerge.

2.3   AN ANALYSIS OF EVOLUTIONARY ECONOMIC DEVELOPMENT

Following the conventional Hirschman (1958) paradigm we assume here that a proper combination of conventional **productive resources** and **public overhead capital** (including R&D) is a necessary condition for balanced growth. These factors are essentially the propulsive motives and incubators for the process of structural economic developments (see also Rosenberg, 1976). It is plausible that in case of qualitative changes in a nonlinear dynamic production system several shocks and perturbations may emerge (see for interesting illustrations also Allen and Sanglier, 1979; Casetti, 1981; Dendrinos, 1981; and Wilson, 1981). A simple mathematical representation of the driving forces of such a production system can be found in Nijkamp (1983, 1984, 1989). This simplified model was based on a so-called **quasi-production function** (including productive capital, infrastructure and R&D capital as arguments). The dynamics of the system were described by motion equations for productive investments, infrastructure investments and R&D investments. Several constraints (i.e., ceilings) were also added, for instance, due to the existence of capacity limits. In our illustration we will start with a simple dynamic neo-classical production function as the basis for a more formal analysis of growth patterns of an economy. The assumption is made that output is generated by a mix of conventional production factors (capital, labour) and public overhead capital (including R&D capital). Later on we will turn to a more complicated and comprehensive economic system (Section 2.4) and also analyse the stability properties of that system. Here, the following Cobb-Douglas production function will be assumed for our (closed) production system:

$$Y = \alpha \, Q^\beta P^\gamma, \qquad (2.1)$$

with Y, Q and P representing output, conventional production factors and social overhead capital, respectively. The parameters β and γ reflect the production elasticities concerned. It is well known that, if instead of social overhead and R&D capital an exponential growth rate of technological progress would have been included in (2.1), the resulting Cobb-Douglas production function would have been at the same time Harrod-, Hicks- and Solow-neutral, provided the technical change concerned would have been disembodied (see also Rouwendal and Nijkamp, 1989, and Stoneman, 1983).

A production function of type (2.1) may only be a reasonable approximation of the underlying production technology within a range of realistic **floors** and **ceilings** (Y_{min}, Y_{max}). Only in this range the production elasticities are assumed to be strictly positive. It is known from the literature on biological population dynamics (e.g. Pimm, 1982) that the existence of either floors or ceilings may generate fluctuating patterns. This is likely to be relevant also in an economic context. Below the minimum threshold level Y_{min}, the critical mass of the economy may be too small to generate economies of scale and scope, so that then a marginal increase in one of the production factors may have a negligible impact on the net output of production. This situation suggests that an economy needs a minimum endowment with production factors before it reaches a self-sustained growth trajectory (see also McKenzie and Zamagni, 1991).

Furthermore, beyond a certain maximum capacity level Y_{max} of the economy, bottleneck phenomena (congestion, diseconomies of scale or scope, e.g.) - caused by a high geographic or industrial concentration of Q - may again lead to a zero or even negative marginal product of conventional production factors. Any further increase in these production factors may then diminish output, unless this situation of a negative marginal product is compensated and corrected by the implementation of new public overhead and R&D investments (the well-known 'depression trigger' phenomenon in Schumpeterian theory).

It is easily seen that, if model (2.1) is explicitly put in a dynamic form, within the relevant range (Y_{min}, Y_{max}) the changes in output in a certain period of time may be approximated by means of the following discrete time version of (2.1):

$$\Delta Y_t = (\beta_t q_t + \gamma_t p_t) Y_{t-1},$$

with:

$$\Delta Y_t = Y_t - Y_{t-1}$$

and: (2.2)

$$q_t = (Q_t - Q_{t-1})/Q_{t-1}$$

$$p_t = (P_t - P_{t-1})/P_{t-1}$$

Hence q_t and p_t are the rates of growth in conventional production factors and social overhead capital and β_t and γ_t are the respective elasticities of output with respect to these inputs. Such a discrete approximation of a model with a continuous time trajectory is usually valid within the range for which the structure of the economic system is stable, and within this range the system will exhibit a non-cyclical growth. This self-sustained growth path may be drawing to a close because of either external causes (e.g., scarcity of production factors or lack of demand) or internal forces (e.g., emergence of diseconomies of scale and scope leading to negative marginal products).

External factors may drive the system toward an upper limit set by the new constraints concerned. **Internal** factors may lead to perturbations and qualitative changes in systemic behaviour. Suppose for instance, a capacity constraint caused by too high a concentration of capital in a production system. Then each additional increase in productive capital will have a negative impact on output. This implies that the production elasticity has become a negative time-dependent variable. In other words, beyond the capacity limit Y_{max} an auxiliary relationship reflecting a negative marginal product of conventional production factors may be assumed, for instance, of the following form:

$$\beta_t = \beta^*(Y_{max} - \kappa Y_{t-1})/Y_{max} \qquad (2.3)$$

In practice the economy may not move beyond Y_{max}, but equation (2.3) shows that as it approaches Y_{max}, the elasticity of output with respect to conventional production factors decreases at a rate of $-\kappa\beta^*/Y_{max}$. Substitution of (2.3) into (2.2) leads to the following adjusted dynamic production function:

$$\Delta Y_t = v_t(Y_{max} - \kappa Y_{t-1})Y_{t-1}/Y_{max} + \gamma p_t Y_{t-1} \qquad (2.4)$$

where $v_t = \beta^* q_t$. This is seemingly a fairly simple non-stochastic dynamic relationship, but it can be shown that this equation is able to generate - under certain conditions - **unstable and even erratic behaviour** leading to a-periodic fluctuations. It is evident that the evolution of v_t itself is likely to be endogenous. The accumulation of capital and human capital, for example, is a function of the level of real income. Hence $\Delta q_t = a(Y_t)$ and, thus, $\gamma v_t = \beta^* \Delta q_t = \beta^* a(Y_t)$. Similarly, the capacity limit Y_{max} may be affected by investments in social overhead capital, i.e. $\Delta Y_{max} = b(q_t)$.

Combining this with equation (2.4), the following dynamic system emerges:

$$\Delta Y_t = v_t(1 - \kappa Y_{t-1}/Y_{max})Y_{t-1} + \gamma p_t Y_{t-1}$$

(2.5)

$$\Delta v_t = \beta^* a(Y_t)$$

$$\Delta Y_{max} = b(q_t)$$

It is noteworthy that system (2.5) is an example of a Lotka-Volterra type model, which has often been used in recent years to model predator-prey relationships in population dynamics (see also Goh and Jennings, 1977; Jeffries, 1979; Pimm, 1982; and Wilson, 1981). However, for the sake of expository purposes we will abstract here from the positive feedback from output to inputs and return to that issue in a model of endogenous input accumulation and technological change in the next section. Hence here we assume that the rates of change in conventional production factors and social overhead capital are both exogenously given. The endogeneity of Y_{max} does also not affect the property of the model we will focus on and hence for the sake of simplicity we will assume that Y_{max} is fixed over the period of interest. Hence (2.5) reduces to:

$$\Delta Y_t = \beta^* q (1 - \kappa Y_{t-1}/Y_{max})Y_{t-1} + \gamma p Y_{t-1}$$

(2.6)

The model represented by equation (2.6) has a similar structure to nonlinear difference equations studied by May (1974), Li and Yorke (1975) and Yorke and Yorke (1975). Applications in a geographical setting can be found in Brouwer and Nijkamp (1985), Dendrinos and Mullally (1983, 1984) and Nijkamp and Reggiani (1989) among others.

Equation (2.6) is a standard equation from population dynamics. It should be noted that logistic evolutionary patterns may also be approximated by a (slightly more flexible) Ricker curve (see May, 1974). In that case, the exponential specification precludes the generation of negative values for the Y variables in simulation experiments, a situation that may emerge in relation to equation (2.6).

It can be easily seen that there are two steady-states, $\bar{Y} = 0$ and $\bar{Y} = (1 + p/\beta q)/\kappa$. However, the stability of the system out of the steady-state equilibria is a complex issue.

Model (2.6) has some very unusual properties. On the basis of numerical experiments, it was demonstrated by May that this model may exhibit a remarkable spectrum of dynamical behaviour, such as stable equilibrium points, stable cyclic oscillations, stable cycles, and chaotic regimes with a-periodic but bounded fluctuations. Two major elements determine the stability properties of (2.6), viz. the **initial values** of Y_t and the tuning parameters (in our case β^* and q) which affect the **growth rate** for the economic system. Simulation experiments indicated that especially the tuning parameters have a major impact on the emergence of cyclic or a-periodic fluctuations. May has demonstrated that a stable equilibrium may emerge if $0 \leq \beta^* q \leq 2$ (and p = 0); otherwise stable cyclic and unstable fluctuations may be generated. Li and Yorke (1975) have later developed a set of sufficient conditions for the emergence of chaotic behaviour for general continuous difference equations.

Clearly, in our discrete model the potential chaotic behaviour depends on the value of β^* and q. It is easily seen from (2.6) that our dynamic model is essentially nothing else but an expression for the growth rate of output generated by the new technological conditions reflected in the production elasticity β^*. Usually such a relative change is positive but less than or equal to 1. It is thus plausible to stipulate that only in case of drastic or structural changes β^* is larger than 1. Similarly, conventional production factors grow at a rate of a few percent per annum. Hence even if the degree of homogeneity of the Cobb-Douglas production function would be higher than 1, $\beta^* q$ would be relatively small, as in case of a normal evolutionary pattern the relative changes in production factors will not be excessively high. Thus in case of incremental changes it is clear that $\beta^* q \leq 1$, so that then a stable equilibrium is ensured; otherwise many alternative evolutionary patterns of the system concerned may emerge. Consequently, the conclusion may be drawn that - due to the presence of a capacity limit Y_{max} - an economy might in principle exhibit a wide variety of dynamical or even cyclical growth patterns, although in this case the emergence of chaos does not seem to be very likely if we consider only short-term small changes.

The variety of behaviour generated by equation (2.6) can be easily demonstrated by means of two simple simulation experiments. In the first experiment there are in the absence of capacity constraints, constant returns to scale with respect to conventional production factors, i.e. $\beta^* = 1$. These production factors are assumed to grow at a rate of 5 percent per annum (q = 0.05). The elasticity of output with respect to social overhead capital is set at $\gamma = 0.3$, while this input grows at a rate of 3 percent ($p = 0.03$). Finally, output is scaled such that $Y_{max} = 1$, $Y_o = 0.1$ and the parameter representing the congestion and other decreasing returns effects $\kappa = 1.4$. In this case, Figure 2 shows that output growth follows a logistic curve with a long-run static equilibrium at $Y = (1+\gamma p/\beta q)/x = (1+0.3 \times 00.3/(1 \times 0.05))/1.4 = 0.843$.

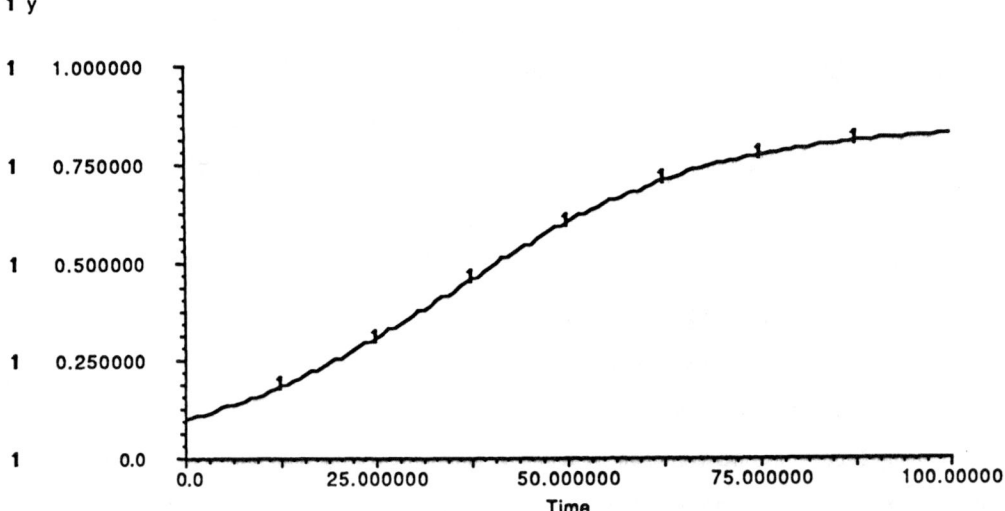

Figure 2.   Results of a simulation run with growth converging to a stationary state.

In the second simulation, we change the time window by considering discrete shocks to productivity occurring every decade. Thus, the time index t refers now to 10 year periods. We also consider much faster growth than in the first case: conventional production factors grow at a rate of 14 percent per annum (e.g. due to a rapid influx of migration and foreign capital). Thus, over the decade the compound growth rate is $q = (1.14)^{10} - 1 = 2.7$. However, social overhead capital continues to grow at a rate of 3 percent p.a., i.e. $p = (1.03)^{10} - 1 = 0.34$. As before, the steady-state equilibrium can be easily computed, viz. $Y = 0.741$, but this equilibrium is now highly unstable. Figure 3 shows that the economy now exhibits wild fluctuations.

It is well known that the outcome of the second simulation run is entirely the consequence of the specification of the model in difference equation form and the choice of the unit of time. In differential equation form, model (2.6) would exhibit global convergence to the long-run steady-state (see also May, 1974). However, economic phenomena often exhibit discontinuities and discrete lags. In this case, a difference equation specification would be quite plausible. Moreover, if the non-linear model contains three or more interacting variables (as is the case in system (2.5)) it may exhibit chaotic patterns and strange attractor sets (rather than a single equilibrium point) even in differential equation form. A well known example is provided by the three-equation Rössler model (1976), which has been applied, for example, in a generic management model by Rasmussen and Mosekilde (1988).

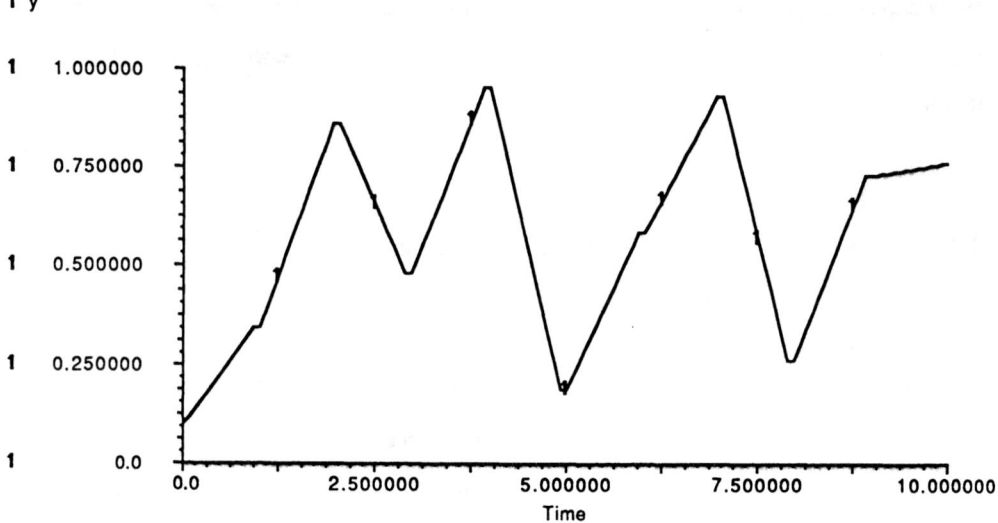

Figure 3.   Results of a simulation run with persistently unstable growth.

In general, the plausibility of the model outcomes would depend on the specification of the model and rigorous empirical scrutiny of its parameters. It must be noted that key parameters (such as the tuning parameter in the May model), which define the qualitative dynamics of the system, may themselves be endogenous and push the system from a chaotic regime to a periodic cycle or stable equilibrium. In this case systems tend to exhibit self-organizing behaviour (see e.g. Radzicki, 1990).

For example, in our case there is a difference with respect to May's model. In May's model, v is a constant, whereas in our case v is endogenously determined by the evolution of our economic system (see equation (2.5)). This has clearly an effect on the growth trajectory, but - given the conditions on v_t - this does not affect the main conclusions regarding the stability of the system, although it has to be realized that drastic changes in any period are likely to generate perturbations in the next period. Since the growth rate v_t is not necessarily a constant, it may become an endogenous variable which may be used as an instrument variable in order to generate a more stable growth path, or to maximise a welfare criterion.

In the latter case an optimal control model emerges. It has been recently discovered that control problems in which there are at least two state variables may generate endogenous and persisting cycles (Feichtinger, 1990) so that even in competitive markets with rational economic agents the system may exhibit persisting, but bounded, fluctuations.

2.4 AN ANALYSIS OF R&D IMPACTS IN CONSTRAINED DEVELOPMENT

Until now we have not explicitly considered the process by which productive inputs accumulate; only their productivity in the presence of rapid growth and capacity constraints was analyzed. In this section we will present a dynamic model of the impact of endogenous R&D in such a constrained economy. Particular attention will again be given to stability properties under conditions of diseconomies of scale.

Research and development (R&D) has become a focal point in current evolutionary economics (cf. Kamien and Schwartz, 1982, Nelson and Winter, 1982, Nijkamp, 1985, and Scherer, 1980). R&D decisions lead, like investment in conventional capital goods, to an interesting intertemporal allocation problem: more R&D expenditures may generate a rise in long-run productivity and profitability, but reduces short-run consumption, and vice versa. This choice problem for capital formation has been extensively studied in traditional economic growth theory, both for economies on a steady state growth path ('golden rule of accumulation') and in the framework of an intertemporal welfare optimisation problem by means of optimal control (see e.g. Ramanathan, 1982).

In the last few years there has been a remarkable revival of interest in economic growth theory (see Barro and Romer, 1990; and Ehrlich, 1990, for overviews). Of particular importance is the role that endogenous technological change can have in the process of development. Such technological change can be the result of human capital accumulation, learning by doing, R&D, innovation diffusion and other forms of spillover effects and spatial interaction (Nijkamp and Poot, 1991).

In general, an important question emerges in relation to R&D and economic growth. The growth path of the economy in an integrated consumption, production, investment and R&D system may be constrained, when the system is facing **capacity limits** (e.g., congestion, other diseconomies of scale, or depletion of exhaustible resources). In Nijkamp et al (1991) the long-run evolutionary path of such an economy was analyzed by means of a dynamic (discrete-time) model incorporating the generation of technological change under conditions of diseconomies of scale. In this section some elements of their approach will be taken up again. It will be shown that a constrained dynamic economic system may generate a wide range of dynamic behaviour, including - for certain parameters - chaotic evolution. We assume that the production in the economy under consideration can be described by means of the following simple production function:

$$Y_t = \epsilon_t K_t, \qquad (2.7)$$

with K_t the installed capital stock at the beginning period t and ϵ_t a time-dependent technological coefficient representing average capital productivity during the period (t,t+1). It should be noted that this linearity assumption is not as restrictive as it seems, since in a sense we may consider (2.7) an identity in which ϵ_t includes all other relevant factors (labour, land, social overhead capital, R&D) which influence capital productivity. Consequently, the elasticity of substitution between capital and other production factors is not assumed to be zero; the time trajectory of ϵ can incorporate both substitution between production factors as well as technological change.

Capital accumulation can be described by means of the following standard expression:

$$K_{t+1} = (1-\delta) K_t + I_t \qquad (2.8)$$

with I_t gross investment during period $(t,t+1)$ and δ the rate of physical depreciation of capital. Now we assume the following simple investment function:

$$I_t = \sigma_t Y_t, \qquad (2.9)$$

with σ_t the average savings rate (assuming the existence of equilibrium between savings and capital increase). The value of s_t will be the outcome of an intertemporal optimisation problem of economic agents in a competitive economy. On a long-run steady-state growth path, v_t would be constant and its value a function of inter alia the discount rate, the technology, the welfare function and population growth.

We take for granted that the current production efficiency can be increased through R&D embodied in the production technology. This requires a change in the production function, as R&D investments will increase efficiency due to a change in the capital coefficient (see Baumol and Wolff, 1984; Mansfield, 1980; Nelson, 1981). In other words, a new 'technological regime' (cf. Nelson and Winter, 1982) requires R&D expenditures with a positive impact on the production efficiency parameter ϵ_t. In our model this effect will be indicated by a parameter v_t, which measures the impact on capital productivity as a result of an additional unit of R&D. This leads to the following equation:

$$\Delta \epsilon_t = \epsilon_{t+1} - \epsilon_t = v_t R_t \qquad (2.10)$$

where R_t represents the R&D investments during period $(t,t+1)$ and v_t the R&D impact parameter for the capital coefficient. Next we may introduce a relationship for R_t, which defines the savings rate for R&D as a proportion of income:

$$R_t = \sigma_2 Y_t \qquad (2.11)$$

Again, the value of σ_2 can be the outcome of an intertemporal optimisation problem. In any period, it is obvious that the amount of output available for consumption C_t is given by:

$$C_t = (1-\sigma_1-\sigma_2) Y_t, \qquad (2.12)$$

Next, substitution of (2.11) into (2.10) leads to the following result:

$$\Delta \epsilon_t = v_{t,\sigma_2} Y_t \qquad (2.13)$$

If v_t were constant over time, capital accumulation would generate ever-increasing growth in output and capital productivity. This highly unlikely outcome suggests that v_t is likely to decrease when output increases. In other words, the marginal efficiency of R&D declines when production increases. Under a given 'technological regime', ultimately a 'saturation' level of output $Y_{t,max}$ is likely to exist at which additional R&D has no longer an impact on productivity. Such a saturation level (ceiling) may arise from capacity limits (technological, social, economic) and reflects - for a given production technology - a 'limits to growth' phenomenon, stemming from diseconomies of scale and scope, as in the previous section. Arguments in favour of the assumption of a decreasing productivity of R&D in case of more mature economic conditions can also be found in Ayres (1987) and Metcalfe (1981) among others.

In view of the above observations it is now clear that $v_t=0$ when $Y_t \geq Y_{t,max}$. Naturally, these limits to growth themselves may be subject to change, so that $Y_{t,max}$ may increase with time and - as prevailing bottlenecks are overcome - new R&D may again have a positive effect on productivity. Thus, the following specification for an adjusted (i.e., time-dependent) R&D impact parameter seems plausible:

$$v_t = \max \{v^* (1-Y_t/Y_{t,max}), 0\} \qquad (2.14)$$

Furthermore, it is plausible that not only would R&D expenditure become ineffective if output expands beyond $Y_{t,max}$, but it may also be expected that diseconomies of scale and scope set in which **reduce** capital productivity. The previous remarks indicate that instead of (2.10) we may now have the following simple relationship for the change in capital productivity:

$$\Delta \epsilon_t = v_t R_t - \mu_t Y_t \tag{2.15}$$

in which μ_t measures the effect of diseconomies on productivity when output exceeds $Y_{t,max}$, so that:

$$\mu_t = \max (\mu^* (Y_t/Y_{t,max} - 1), 0) \tag{2.16}$$

Assuming for simplicity that $Y_{t,max}$ grows at the exogenous rate n and substituting (2.16) and (2.14) into (2.10), the **motion** in the system can now be described by the following set of nonlinear difference equations;

$$K_{t+1} = (1-\delta) K_t + \sigma_1 \epsilon_t K_t$$

$$\epsilon_{t+1} = \epsilon_t + [\sigma_2 v^* \max (1-Y_t/Y_{t,max}, 0) -$$

$$- \mu^* \max (Y_t/Y_{t,max} - 1, 0)] \epsilon_t K_t \tag{2.17}$$

$$Y_{t+1} = \epsilon_{t+1} K_{t+1}$$

$$Y_{t+1, max} = (1+n)^t Y_{0,max}$$

In view of the nonlinear properties of this model, it is clear that for any given initialisation $(K_o, \epsilon_o, Y_o, Y_{o,max})$ system (2.17) can exhibit a wide range of time trajectories dependent on parameter values.

Some results based on simulation experiments are illustrated in Figures 4 and 5. Figure

4 is based on the assumption that $K_o=1000$ and the capital-output ratio equals 5; hence $\epsilon_o=0.2$ and $Y_o=200$. The saving ratio is 20 percent; 2 percent of the capital stock is assumed to become obsolete each period and 2 percent of income is spent on R&D, so that $\sigma_1=0.20$ and $\delta=\sigma_2=0.02$. The **sustainable** output capacity $Y_{o,max}=1000$ and grows at 1 percent. Moreover, $\mu^*=0.0001$ and $v^*=0.001$. Since $5\sigma_{-v}^*=\mu^*$, the productivity response is five times as elastic when $Y_t > Y_{t,max}$ than when $Y_t < Y_{t,max}$, and of opposite sign.

Figure 4 shows that growth in the system is - under these conditions - in initial periods **accelerating**. However, the growth rate of capital productivity reaches a maximum at $t=25$ and subsequently declines, until Y_t reaches the saturation level $Y_{t,max}$ at $t=37$. At this point, the growth rate of capital accumulation reaches a maximum. Beyond $t=37$, Y_t will remain above $Y_{t,max}$ but it will converge to the latter value. Consequently, capital productivity becomes constant at a rate of $(n+\delta)/\sigma_1=0.15$, whilst capital and output grow at a steady-state rate of 1 percent.

Next, we assume that in Figure 5 all parameters are the same as in Figure 4, but μ^* has been increased to five times its former value. Consequently, the effect of diseconomies is now sufficiently strong to push Y_t at times below $Y_{t,max}$ so that **growth cycles** are generated with a variable **periodicity** but with decreasing amplitude. The system eventually converges again to a steady-state growth of 1 percent. Thus stringent diseconomies cause the system to be more chaotic.

The earlier noted sensitivity of models which exhibit chaotic behaviour to parameter values or initial conditions can be easily demonstrated by means of the model discussed in this section. Figure 6 duplicates the fluctuations in the income growth rate displayed in Figure 5, but the growth rates of capital and capital productivity have been deleted for clarity, while the focus is on period 60 to 100 only. Figure 6 shows the outcome of simulation with exactly the same model, but with parameter μ increased by 1 percent (i.e. from 0.0005 to 0.000505). This very small change in the parameter value generates nonetheless fluctuations in the growth rate of income which, for a large proportion of the time, are very different from those for the earlier simulation. Hence even if the model which generated Figure 6 would be known perfectly, except for the exact value of one of the parameters, such as μ, it would still be impossible to forecast the level of income for any period but the very near future.

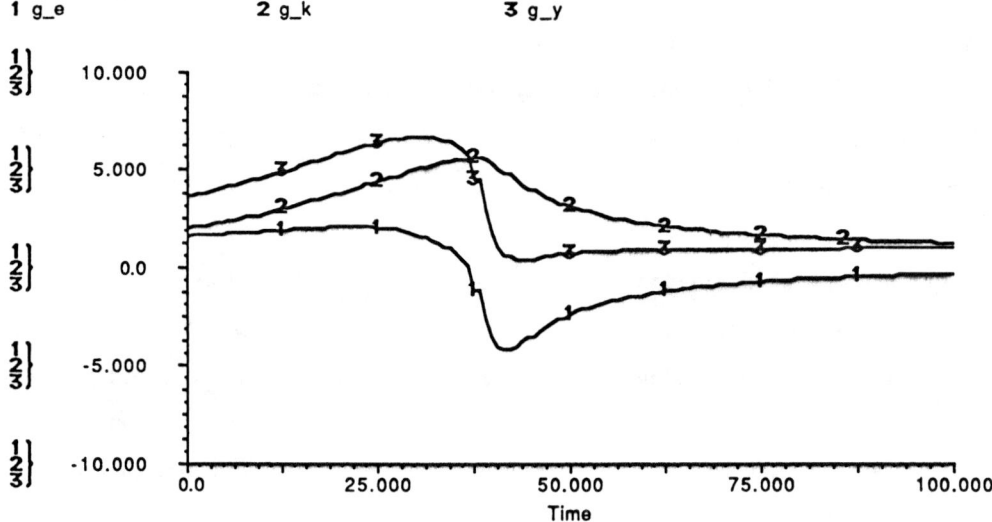

Figure 4. Growth converging to a steady state.
Legend: 1 : growth in capital productivity
         2 : capital growth rate
         3 : income growth rate

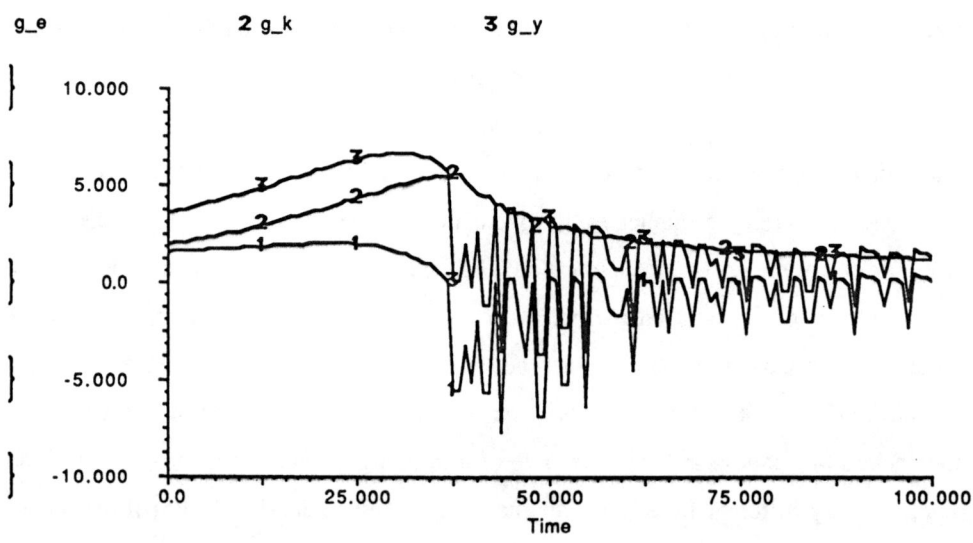

Figure 5. Growth cycles generated by strong external diseconomies.

Similarly, a small change in the initialisation of the system can generate after some time drastic qualitative and quantitative changes in the time trajectories of the variables of the system. It should be noted that such a sensitive dependence to initial conditions and parameter values applies also to unstable linear systems, but in the non-linear case the resulting time trajectories may remain bounded, while in the linear case any divergence will be monotic.

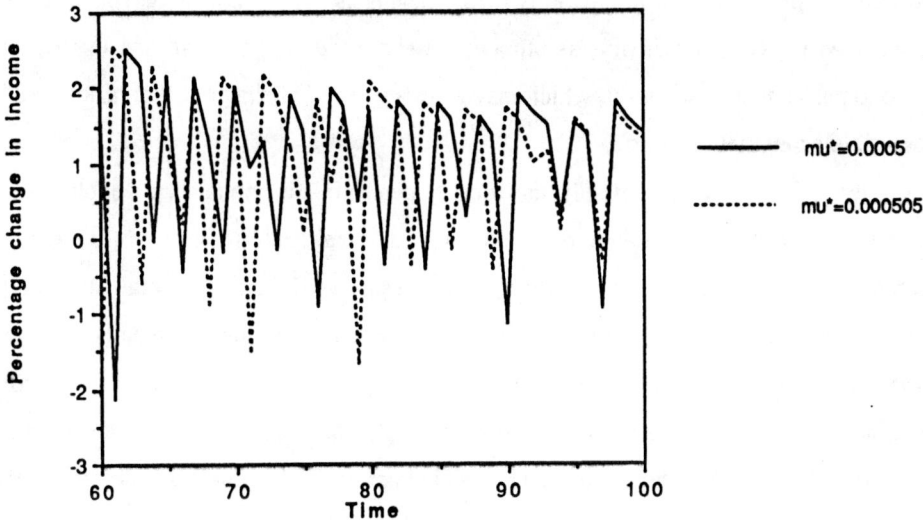

Figure 6. The high sensitivity of the non-linear growth model to parameter changes.

2.5 RETROSPECT

The previous experiments have demonstrated that even conventional economic models may exhibit irregular behaviour in case of high growth rates or strict limits to growth. In various cases this may be beyond plausible empirical values of an economic system, so that chaos in a real-world system is less likely to emerge. However, in case of rapid transitions or sudden adjustments such chaos patterns may temporarily emerge.

If instead of a simulation experiment, one would have to use models of the above nature as normative policy models, it would be necessary to introduce an appropriate objective (or welfare) function encompassing a trade-off between relevant welfare arguments. A dynamic programming or optimal control formulation would then be desirable (see Kendrick, 1981, and Nijkamp and Reggiani, 1988, 1989). Such a constrained dynamic optimization might in principle reduce chaotic fluctuations inherent in the nonlinear dynamics of the growth model and provide a self-organising stabiliser for the system.

Another point concerns the specification of nonlinear dynamic models. It may be important to stress that the foundations of specifying an economic model have to be firmly rooted in economic theory, as otherwise we run the danger of ad hoc theorizing and econometric misspecification, which may generate chaotic behaviour that is not based on plausible economic grounds.

However, it is extremely unlikely that purely deterministic models will explain the outcomes of interaction between economic agents. The central question for future research is whether observed volatility is the result of linear structural dynamics combined with exogenous shocks and stochastic noise due to measurement error, or alternatively, deterministic nonlinear dynamics in which stochastic terms play a minor role. Since the number of data points in economic phenomena is so much smaller than in, for example, the experimental sciences in which nonlinear dynamics is becoming very popular, nonlinear determinism and the presence of a large number of exogenous shocks are observationally equivalent.

Finally, it is important to call attention to the fact that in various cases a system is not chaotic as a whole, but has only a few 'niches' (modules or equations) which under certain conditions may exhibit chaotic behaviour. The question whether chaotic behaviour of a small sub-system will be dampened by the dominance of another and otherwise stable system, or whether it will exert an explosive influence upon a whole system needs further investigation.

Although nonlinear models for economic development may provide an interesting explanatory framework for the rise and decline of regions and nations in a dynamic (sometimes chaotic) context, it is also evident from the above experiments that the 'economics of chaos' desperately needs rigorous empirical research work. The challenge is to build theoretical models which combine small amounts of randomness with nonlinearities and succeed in generating data that replicate real economic and financial time series.

REFERENCES

Albin, P.S., 1987, Micreconomic Foundations of Cyclical Irregularities or Chaos, **Mathematical Social Sciences**, vol. 13, pp. 185-214.

Allen, A., 1989, The New Zealand Sharemarket Gross Index and Chaos: Some preliminary evidence, Discussion Paper in Money and Finance, Faculty of Commerce and Administration, Victoria University.

Allen, P.M. and M. Sanglier, 1979, A Dynamic Model of Growth in a Central Place System, **Geographical Analysis**, vol. 11, pp. 256-272.

Andersen, D.F. (ed.), 1988, **Chaos in System Dynamic Models: Special Issue System Dynamics Review**, vol. 4, no. 1/2.

Ayres, R.U., 1987, Barriers and Breakthroughs: An "Expanding Frontiers" Model of the Technology-Industry Life Cycle, Research Paper, IIASA, Laxenburg.

Balducci, R., G. Candela and G. Ricci, 1987, A Generalization of R. Goodwin's Model with Rational Behaviour of Economic Agents, **Nonlinear Models of Fluctuating Growth**, (R.M. Goodwin, M. Krüger and A. Vercelli, eds.), Springer-Verlag, Berlin, pp. 47-66.

Barnett, W. and P. Chen, 1988, The Aggregation-theoretic Monetary Aggregates Are Chaotic and Have Strange Attractors, **Dynamic Econometric Modelling**, (W. Barnett, E. Berndt and H. White, eds.), Cambridge University Press, pp. 199-245.

Barnett, W., J. Geveke, and K. Shell (eds.), 1990, **Economic Complexity: Chaos, Sunspots, Bubbles and Nonlinearity**, Cambridge University Press, Cambridge.

Barro, R.J. and P.M. Romer, 1990, Economic Growth, **NBER Reporter**, National Bureau of Economic Research, Fall 1990, pp. 1-5.

Batty, M., 1991, Cities as Fractals: Simulating Growth and Form, **Fractals and Chaos** (A.J. Crilly, R.A. Earnshaw, and H. Jones, eds.), Springer Verlag, Berlin, pp. 43-70.

Baumol, W.J. and J. Benhabib, 1989, Chaos: Significance, Mechanism, and Economic Applications, **Journal of Economic Perspectives**, vol. 3, no. 1, pp. 77-105.

Baumol, W.J. and R.E. Quandt, 1985, Chaos Models and their Implications for Forecasting, **Eastern Economic Journal**, vol. 11, no.1, pp. 3-15.

Baumol, W.J., and E.N. Wolff, 1983, Feedback from Productivity Growth in R&D, **Scandinavian Journal of Economics**, vol. 85, pp. 145-157.

Baumol, W.J., and E.N. Wolff, 1984, Feedback Models: R&D Information, and Productivity Growth, **Communication and Information Economics**, (M. Jussawalla and H. Ebenfield, eds.), North-Holland Publ. Co., Amsterdam, pp. 73-92.

Benhabib, J. and R.H. Day, 1981, Rational Choice and Erratic Behaviour, **Review of Economic Studies**, vol. 48, pp. 459-471.

Benhabib, J. and R.H. Day, 1982, A Characterization of Erratic Dynamics in the Overlapping Generations Model, **Journal of Economic Dynamics and Control**, vol. 4, pp. 37-55.

Blommestein, H.J., 1986, **Eliminative Econometrics**, Ph.D. Diss., Department of Economics, Free University, Amsterdam.

Boldrin, M., 1988, Persistent Oscillations and Chaos in Dynamic Economic Models: Notes for a Survey, **SFI Studies in the Science of Complexity**, Addison-Wesley Publ. Co., pp. 49-75.

Boldrin, M., and M. Woodford, 1990, Equilibrium Models Displaying Endogenous Fluctuations and Chaos: A Survey, **Journal of Monetary Economics**, vol. 25, pp. 189-222.

Brock, W.A., 1989, Chaos and Complexity in Economic and Financial Science, in G.M. von Furstenberg (ed.), **Acting under Uncertainty: Multidisciplinary Conceptions**, Kluwer, Norwell, MA.

Brock, W.A., 1988, Applications of Nonlinear Science Statistical Inference Theory to Finance and Economics. Proceedings of the Spring Seminar of the Institute for Quantitative Research in Finance, Innisbrook, Florida.

Brock, W.A., 1986, Distinguishing Random and Deterministic Systems, **Journal of Economic Theory**, vol. 40, pp. 168-195.

Brock, W.A., Dechert, W.D. and J.A. Scheinkman, 1987, A Test for Independence Based on the Correlation Dimension, SSRI Working Paper No. 8702, Department of Economics, University of Wisconsin at Madison.

Brock, W.A. and C.L. Sayers, 1988, Is the Business Cycle Characterized by Deterministic Chaos? **Journal of Monetary Economics**, vol. 22, pp. 71-90.

Brouwer, F. and P. Nijkamp, 1985, Qualitative Structure Analysis of Complex Systems, **Measuring the Unmeasurable; Analysis of Qualitative Spatial Data** (P. Nijkamp, H. Leitner and N. Wrigley, eds.), Martinus Nijhoff, The Hague, pp. 509-532.

Casetti, E., 1981, Technological Progress, Exploitation and Spatial Economic Growth: A Catastrophe Model, **Dynamic Spatial Models** (D.A. Griffith and R. MacKinnon, eds.), Sijthoff and Noordhoff, Alphen aan de Rijn, pp. 215-277.

Chen, P., 1988, Empirical and Theoretical Evidence of Economic Chaos, **System Dynamics Review**, vol. 4, no. 1, pp. 81-108.

Chiarella, C., 1988, The Cobweb Model: Its Instability and the Onset of Chaos, **Economic Modelling**, vol. 5, no. 4, pp. 377-384.

Crilly, A.J., R.A. Earnshaw, and H. Jones (eds.), 1990, **Fractals and Chaos**, Springer Verlag, Berlin.

Crutchfield et al., 1982, Fluctuations and Simple Chaotic Dynamics, **Physics Reports**, vol. 92, pp. 45-82.

Dana, R.A. and L. Montrucchio, 1986, Dynamic Complexity in Duopoly Games, **Journal of Economic Theory**, vol. 40, pp. 40-56.

Day, R.H., 1982, Irregular Growth Cycles, **American Economic Review**, vol. 72, pp. 406-414.

Dendrinos, D.S. (ed.), 1981, **Dynamic Non-Linear Theory and General Urban/Regional Systems**, School of Architecture and Urban Design, Lawrence, Kansas.

Dendrinos, D.S., 1984, Turbulence and Fundamental Urban/Regional Dynamics, Paper presented at the American Association of Geographers, Washington D.C., April 1984.

Dendrinos, D.S., 1986, On the Incongruous Spatial Employment Dynamics, **Technological Change, Employment and Spatial Dynamics**, (P. Nijkamp, ed.), Springer-Verlag, Berlin, pp. 321-339.

Dendrinos. D.S. and H. Mullally, 1983, Empirical Evidence of Volterra-Lotka Dynamics in United States Metropolitan Areas: 1940-1977, **Evolving Geographical Structures** (D.A. Griffith and T. Lea, eds.), Martinus Nijhoff, The Hague, pp. 170-195.

Dendrinos. D.S. and H. Mullally, 1984, **Urban Evolution**, Oxford University Press, Cambridge.

Dendrinos, D.S. and M. Sonis, 1987, The Onset of Turbulence in Discrete Relative Multiple Spatial Dynamics, **Applied Mathematics and Computation**, vol. 22, pp. 25-44.

Deneckere, R. and S. Pelikan, 1986, Competitive Chaos, **Journal of Economic Theory**, vol. 40, pp. 13-25.

Devaney, R.L., 1986, **Chaotic Dynamical Systems**, Benjamin Cummings Publ. Co., Menlo Park, CA.

Ehrlich, I., 1990, The Problem of Development: Introduction, **Journal of Political Economy**, vol. 98, no. 5.2, pp. S1-S11.

Fama, E., 1976, **Foundations of Finance**, Basic Books, New York.

Feichtinger, G., 1990, Optimal Periodic Policies in Dynamic Economic Systems, University of Technology, Vienna, Austria, mimeo.

Finkenstädt, B. and P. Kuhbier, 1990, Chaotic Dynamics in Agricultural Markets, Free University of Berlin, mimeo.

Funke, M., 1987, A Generalized Goodwin Model Incorporating Technical Progress and Variable Prices, **Economic Notes**, vol. 2, pp. 36-47.

Goh, B.S. and L.S. Jennings, 1977, Feasibility and Stability in Randomly Assembled Lotka-Volterra Models, **Ecological Modelling**, vol. 3, no. 1, pp. 63-71.

Goodwin, R.M., 1982, **Essays in Economic Dynamics**, Macmillan, London.

Grandmont, J.M., 1985, On Endogenous Competitive Business Cycles, **Econometrica**, vol. 53, pp. 995-1046.

Grandmont, J.M. (ed.), 1986, **Special Issue Journal of Economic Theory**, vol. 40.

Guckenheimer, J. and P. Holmes, 1983, **Non-Linear Oscillations, Dynamical System and Bifurcation of Vector Fields**, Springer Verlag, Berlin.

Guckenheimer, J., G. Oster and A. Ipatchi, 1977, The Dynamics of the Density Dependent Population Models, **Journal of Mathematical Biology**, vol. 4, pp. 101-147.

Haag, G. and W. Weidlich, 1983, A Nonlinear Dynamic Model for the Migration of Human Population, **Evolving Geographical Structures** (D.A. Griffith and T. Lea, eds.), Martinus Nijhoff, The Hague, pp. 24-61.

Hirschman, A.O., 1958, **The Strategy of Economic Development**, Yale University Press, New Haven.

Hommes, C.H. and H.E. Nusse, 1989, Does an Unstable Keynesian Unemployment Equilibrium in a Non-Walrasian Dynamic Macroeconomic Model Imply Chaos?, **Scandinavian Journal of Economics**, vol. 91, pp. 161-167.

Hommes, C.H., H.E. Nusse and A. Simonovits, 1990, Hicksian Cycles and Chaos in a Socialilist Economy, Research Memorandum, Institute for Economic Research, State University, Groningen.

Hsieh, D., 1989, Testing for Nonlinear Dependence in Daily Foreign Exchange Rates, **Journal of Business**, vol. 62, pp. 339-368.

Jeffries, C., 1979, Qualitative Stability and Digraphs in Model Ecosystems, **Ecology**, vol. 55, no. 6, pp. 1415-1419.

Kamien, M. and N. Schwartz, 1982, **Market Structure and Innovations**, Cambridge University Press, Cambridge.

Kelsey, D., The Economics of Chaos or the Chaos of Economics, **Oxford Economic Papers**, 40, 1988, pp. 1-31.

Kendrick, D., 1981, Control Theory with Applications to Economics, **Handbook of Mathematical Economics** (K.J. Arrow and M.D. Intriligator, eds.), North-Holland Publ. Co., Amsterdam, pp. 111-158.

Kleinknecht, A., 1986, **Innovation Patterns in Crisis and Prosperity; Schumpeter's Long Cycle Reconsidered**, MacMillan, London.

Lasota, A. and M.C. Mackey, 1985, **Probabilistic Properties of Deterministic Systems**, Cambridge University Press, Cambridge.

LeBaron, B., 1989, Diagnosing and Simulating Some Asymmetries in Stock Return Volatility, Dept. of Economics, University of Wisconsin-Madison, mimeographed paper.

Li, T. and J.A. Yorke, 1975, Period Three Implies Chaos, **American Mathematical Monthly**, vol. 82, pp. 985-992.

Lichtenberg, A.J. and M.A. Lieberman, 1983, **Regular and Stochastic Motion**, Springer Verlag, Berlin.

Liosattos, P., 1980, Spatial Dynamics: Some Conceptual and Mathematical Issues, **Environment & Planning A**, vol. 12, pp. 1051-1071.

Lorenz, H.W., 1987, International Trade and the Possible Occurrence of Chaos, **Economics Letters**, vol. 23, pp. 135-138.

Lorenz, H.W., 1989, **Nonlinear Dynamical Economics and Chaotic Motion**, Springer Verlag, Berlin.

Lung, Y., 1988, Complexity and Spatial Dynamics Modelling: From Catastrophic Theory to Self-Organizing Process: a Review of Literature, **The Annals of Regional Science**, vol. 22, no. 2, pp. 81-111.

Mansfield, E., 1980, **Industrial Research and Technological Innovation**, W.W. Norton, New York.

May, R.M., 1974, Biological Populations with Non-overlapping Generations, **Science**, no. 186, pp. 645-647.

May, R.M., 1976, Simple Mathematical Models with Very Complicated Dynamics, **Nature**, vol. 261, pp. 459-467.

McKenzie, L.W. and S. Zamagni (eds.), 1991, **Value and Capital: Fifty Years After**, MacMillan, London.

Metcalfe, J.S., 1981, Impulse and Diffusion in the Study of Technical Change, **Futures**, vol. 13, pp. 347-359.

Nelson R., 1981, Research and Productivity Growth and Productivity Differences, **Journal of Economic Literature**, vol. 8, pp. 1029-1064.

Nelson, R.D. and S.G. Winter, 1982, **An Evolutionary Theory of Economic Change**, Harvard University Press, Cambridge.

Nijkamp, P., 1983, Technological Change, Policy Response and Spatial Dynamics, **Evolving Geographical Structures** (D.A. Griffith and T. Lea, eds.), Martinus Nijhoff, The Hague, pp. 75-99.

Nijkamp, P., 1984, A Multidimensional Analysis of Regional Infrastructure and Economic Development, **Regional and Industrial Development Theories, Models and Empirical Evidence** (A.E. Andersson, W. Isard and T. Puu, eds.), North-Holland Publ. Co., Amsterdam, pp. 267-293.

Nijkamp, P., 1985, **Employment, Technological Change and Spatial Dynamics**, Springer Verlag, Berlin.

Nijkamp, P., 1987, Long-Term Economic Fluctuations: A Spatial View, **Socio-Economic Planning Sciences**, vol. 21, no. 3, pp. 189-197.

Nijkamp, P., 1989, Evolutionary Patterns of Urban Production Systems, Belgian Journal of Operations Research, **Statistics and Computer Science**, vol. 29, no. 2, pp. 23-40.

Nijkamp, P. and J. Poot, 1991, Endogenous Technological Change and Spatial Interdependence, Dept. of Economics, Free University, Amsterdam, mimeo.

Nijkamp, P., J. Poot, and J. Rouwendal, 1991, A Non-linear Dynamic Model of Spatial Economic Development and R&D Policy, **Annals of Regional Science**, forthcoming.

Nijkamp, P. and A. Reggiani, 1998, Dynamic Spatial Interaction Models: New Directions, **Environment and Planning A**, vol. 20, pp. 1449-1460.

Nijkamp, P. and A. Reggiani, 1989, Theory of Chaos: Relevance for Analysing Spatial Processes, **Spatial Choices and Processes**, (M.M. Fischer, P. Nijkamp, and Y. Papageorgiou, eds.), North-Holland Publ. Co., Amsterdam, pp. 49-80.

Nijkamp, P. and A. Reggiani, 1990, Logit Models and Chaotic Behaviour, **Environment and Planning A**, vol. 22, pp. 1455-1467.

Nijkamp, P., and A. Reggiani, 1992, **Interaction, Evolution and Chaos in Space**,

Springer Verlag, Berlin.

Nusse, H.E. and C.H. Hommes, 1990, Resolution of Chaos with Application to a Modified Samuelson Model, **Journal of Economic Dynamics and Control**, vol. 14, pp. 1-19.

Pimm, S.L., 1992, **Food Webs**, Chapman and Hall, London.

Pohjola, M.T., 1981, Stable and Chaotic Growth: the Dynamics of a Discrete Version of Goodwin's Growth Cycle Model, **Zeitschrift für Nationalökonomie**, vol. 41, pp. 27-38.

Priestley, M.B., 1988, **Nonlinear and Non-Stationary Time Series Analysis**, Academic Press, London.

Prigogine, I. and I. Stengers, 1985, **Order out of Chaos**, Fontana, London.

Prskawetz, A., 1990, Are Demographic Indicators Deterministic or Stochastic?. Institute for OR, Vienna Technical University, mimeo.

Radzicki, M.J., 1990, Institutional Dynamics, Deterministic Chaos and Self-Organizing Systems, **Journal of Economic Issues**, vol. 24, pp. 57-102.

Ramanathan, R., 1982, **Introduction to the Theory of Economic Growth**, Springer Verlag, Berlin.

Ramsey, J.P., C.L. Sayers and P. Rothman, 1990, The Statistical Properties of Dimension Calculations Using Small Data Sets: Some Economic Applications, **International Economic Review**, vol. 31, pp. 991-1020.

Rand, D., 1978, Exotic Phenomena in Games and Duopoly Models, **Journal of Mathematical Economics**, vol. 5, pp. 173-184.

Rasmussen, D.R. and E. Mosekilde, 1988, Bifurcations and Chaos in a Generic Management Model, **European Journal of Operational Research**, vol. 35, pp. 80-88.

Rasmussen, D.R., E. Mosekilde and J.D. Sterman, 1985, Bifurcations and Chaotic Behavior in a Simple Model of the Economic Long Wave, **System Dynamics Review**, vol. 1, pp. 92-110.

Reggiani, A., 1990, **Spatial Interaction Models: New Directions**, Ph.D. Dissertation, Dept. of Economics, Free University, Amsterdam.

Reiner, R., M. Munz, G. Haag and W. Weidlich, 1986, Chaotic Evolution of Migration Systems, **Sistemi Urbani**, vol. 2/3. pp. 285-308.

Ricci, G., 1985, A Differential Game of Capitalism, **Optimal Control Theory and**

Economic Analysis (G. Feichtinger, ed.), Elsevier, Amsterdam, pp. 633-643.

Rosenberg, N., 1976, **Perspective on Technology**, Cambridge University Press, Cambridge.

Rosser, J.B., 1991, **From Catastrophe to Chaos: A General Theory of Economic Discontinuities**, Kluwer, Dordrecht.

Rössler, O.E., 1976, An Equation for Continuous Chaos, **Physics Letters**, 57A, July 1976, pp. 397-398.

Rouwendal, J., and P. Nijkamp, 1989, Endogenous Production of R&D and Stable Economic Development, **De Economist**, vol. 137, no. 2, pp. 202-216.

Scheinkman, J. A., 1990, Nonlinearities in Economic Dynamics, **The Economic Journal**, vol. 100, Conference 1990, pp. 33-48.

Scheinkman, J. A. and B. LeBaron, 1989, Nonlinear Dynamics and Stock Returns, **Journal of Business**, July 1989, pp. 311-337.

Scherer, F.M., 1980, **Industrial Market Structure and Economic Performance**, Rand McNally, Chicago.

Schuster, H.G., 1984, **Deterministic Chaos**, Physik Verlag, Weinheim, FRG.

Sonis, M., 1986, A Unified Theory of Innovation Diffusion, Dynamic Choice of Alternatives, Ecological Dynamics and Urban/Regional Growth and Decline, **Richerche Economiche**, vol. XL, 4, pp. 696-723.

Sonis, M., 1987, Discrete Time Choice Models Arising from Innovation Diffusion Dynamics, **Sistemi Urbani**, no. 1, pp. 93-107.

Sterman, J.D., 1985, A Behavioural Model of the Economic Long Wave, **Journal of Economic Behaviour and Organization**, vol. 5, pp. 17-53.

Stewart, I., 1989, **Does God Play Dice?**, Basil Blackwell, Oxford.

Stoneman, P., 1983, **The Economic Analysis of Technological Change**, Oxford University Press, Oxford.

Stutzer, M., 1980, Chaotic Dynamics and Bifurcation in a Macro-model, **Journal of Economic Dynamics and Control**, vol. 2, pp. 253-276.

Turner, J., 1980, Non-equilibrium Thermodynamics, Dissipative Structures and Self-Organization, **Dissipative Structures and Spatio-temporal Organization Studies in Biomedical Research** (G. Scott and J. MacMillan, eds.), Iowa State University Press, Iowa, pp. 13-52.

White, R., 1985, Transition to Chaos with Increasing System Complexity: The Case of Regional Industrial Systems, **Environment and Planning A**, vol. 17, pp. 387-396.

Wilson, A.G., 1981, **Catastrophe Theory and Bifurcation**, Croom Helm, London.

Yorke, J.A., and E.D. Yorke, 1981, Chaotic Behaviour and Fluid Dynamics. **Hydrodynamic Instabilities and the Transition to Turbulence** (H.L. Swinney and J.P. Collub, eds.), Springer Verlag, New York, pp. 112-128.

CHAPTER 3
SURVEY OF NONLINEAR DYNAMIC MODELLING IN ECONOMICS
J. Barkley Rosser, Jr.[1]

3.1   INTRODUCTION

This paper will survey recent developments in both the econometrics and the theory of nonlinear dynamic modelling in economics. In contrast to the usual order we shall consider econometric techniques first, partly because many of these are somewhat atheoretical in terms of any associated economic theory. To the extent that there is associated theory it will be noted in passing with the econometric discussion.

The theoretical section will focus on recent developments, some of which may be difficult to pin down econometrically. The major criterion for inclusion is that there be potential applications in the area of nonlinear dynamic spatial modelling with an evolutionary orientation. Some effort will be made to suggest which, if any, of the previously discussed econometric techniques might be most appropriate for dealing with these models.

One point that can be made here is that generally the time scales involved in serious nonlinear spatial models are substantially longer than those in the time series usually studied by most nonlinear econometric techniques. Indeed one of the most important questions in nonlinear dynamic spatial modelling is precisely the interrelation between processes operating at different time scales.

3.2   ECONOMETRICS OF NONLINEAR TIME SERIES ANALYSIS

A.   What is a Nonlinear Time Series?

There is much discussion of nonlinear time series and dynamics in economics. But it turns out that what actually constitutes a nonlinear time series or dynamic process is not exactly obvious on the surface. We follow Priestly (1988) in defining $\{X_t\}$ as a stochastic linear process if it is strictly stationary[2] and can be represented by

$$X_t = \sum_{j=1}^{\infty} a_j N_{t-j} = a(B) N_t, \; a_0 = 1 \text{ and } B^j N_t = N_{t-j} \tag{3.1}$$

where $\{N_t\}$ is an independent and identically distributed (iid) stochastic process and a(B) is a rational polynomial in B.[3] This implies that X_t can be fully represented by an ARMA process based on polynomial operators of a(B) (Brock and Potter, 1991).

This is reasonably straightforward. The problem arises from a theorem of Wold (1938) that states that **any** stationary process can be expressed as a linear system generating uncorrelated impulses known as a Wold representation. This would seem to render uninteresting any effort at nonlinear stochastic modelling. LeBaron (1991) lists three reasons why this is not the case.

One is that the Wold representation necessary to represent an underlying nonlinear process may be extremely complicated. This arises from the fact that although disturbances may be uncorrelated they are not necessarily independent. A nonlinear model may account for these interdependencies more simply.

Second is that the Wold representation will not accurately model the higher moments of the distribution, most notably the variance. A major area of nonlinear modelling in economics has emphasized exactly this issue in the form of allowing variance to conditionally vary in the form of autoregressive conditional heteroskedasticity (ARCH), developed initially by Engle (1982).

Finally, and most important from the perspective of this paper, is that important elements of the qualitative behavior of a nonlinear system will not be captured by the Wold representation. Potter (1991a) has used the Volterra expansion[4] to analyze nonlinear impulse response functions to show that the Wold representation fails to capture the nature of the persistence of the response to shocks to a non linear system. Thus nonlinear modelling is still of interest despite the Wold theorem.

B. Measures of Chaos and Related Concepts

Perhaps the most dramatic of nonlinear models are those that generate deterministic chaos. Much of the interest arises from the fact that such models appear to behave in a completely random way even though they may be driven by deterministic dynamics. The interest was further heightened after it became known that reasonable theoretical nonlinear economic models could generate such chaotic dynamics.[5]

A curious but not widely noted fact about the discussion of chaos is that there remains an unresolved debate regarding what it really is. Probably the most widely used definition is that a system possess sensitive dependence on initial conditions (SDIC). This definition has been especially emphasized by Eckmann and Ruelle (1985) and is associated with a dynamical system possessing at least one positive Lyapunov exponent. These are formally defined by the following. If F is the dynamical function, $F^t(x)$ is the t-th iterate of F starting at initial condition x, D is the derivative, and v is a direction vector, then Lyapunov exponents are solutions to

$$L = \lim_{t \to \infty} (\ln(\|DF^t(x) - \vec{v}\|)/t). \tag{3.2}$$

Despite the emphasis given to this definition of chaotic dynamics little work has been done on estimating such exponents in economics. Algorithms have been developed by Wolf, Swift, Swinney, and Vastano (1985) and by Eckmann, Kamphorst, Ruelle, and Ciliberto (1986). Eckmann, Kamphorst, Ruelle, and Scheinkman (1988) tentatively found positive Lyapunov exponents for stock return series as have Chavas and Holt (1991) for the pork cycle and the milk cycle (Chavas and Holt, 1992). But there does not yet exist a theory of statistical inference for such estimates.[6]

An alternative definition of chaotic dynamics is that a system possess a strange attractor, precisely defined as possessing a homoclinic orbit (Guckenheimer and Holmes, 1983). Strange attractors are characterized by possessing non-integer (or fractal) dimensionality. One approach to estimating dimensionality was developed by Grassberger and Procaccia (1983) in the form of the correlation dimension given by

$$c_m(r) = \lim_{t \to \infty} \ln N(r) / \ln r, \tag{3.3}$$

where N(r) is the number of points whose distance from each other is less than r. In this form the lower the correlation dimension the more chaotic is the system. This measure has an information interpretation also; the lower the dimension the more deterministic the system is.

Because of certain difficulties with this measure Brock, Dechert, and Scheinkman (1987) developed the BDS statistic given by

$$BDS(m,r) = N^{1/2}\{C_m(r) - [C_1(r)^m]\}/b_m, \quad m > 1, \qquad (3.4)$$

where m is an embedding dimension, b_m is the standard deviation, and N is the length of the time series. A test for nonlinearity involves rejecting the null hypothesis that this statistic equals zero for the residuals of a linear representation of a time series.[7] This statistic has been used to test for nonlinearity of numerous economic series. Many have exhibited such nonlinearity, an incomplete list including work stoppages (Sayers, 1988), pig iron production (Brock and Sayers, 1988), US stock returns (Scheinkman and LeBaron, 1989), gold and silver markets (Frank and Stengos, 1989), foreign exchange rates (Hsieh, 1989), pork production (Chavas and Holt, 1991), and milk production (Chavas and Holt, 1992).

Despite these suggestive results it is argued by most observers that deterministic chaos has not been demonstrated for any economic time series. This is based on arguing that dimensionality should be maintained for different measurement intervals and also that out-of-sample forecasts using nearest neighbor techniques[8] should be superior to a random walk as measured by root mean square errors. For US stock returns, one of the more promising series, the former was rejected by Mayfield and Mizrach (1989) and the latter was rejected by Hsieh (1991). Chavas and Holt (1991) maintain the possibility of chaotic dynamics for the pork cycle and also for the milk cycle (1992), but have not carried out the above tests because of the long periods involved in their models.

That these animal products might be possible candidates is due to the existence of production lags that allow for possible cobweb effects. Artstein (1983) and Jensen and Urban (1984) showed possible chaotic dynamics for cobwebs with non-monotonic supply curves, a result generalized to the Walrasian tâtonnement process by Day and Piangiani

(1991), and Chiarella (1988) has shown possible chaotic dynamics even for monotonic supply and demand curves if there are expectation and production lags. Furthermore in the form of the corn-hog cycle pork production can be viewed as part of a predator-prey system, for which the possibility of chaotic dynamics was first shown by May (1974). Although such systems usually depend on discrete time periods, Invernizzi and Medio (1991) have shown the possibility of chaotic dynamics in continuous time models with appropriately complicated lag structures.

Chavas and Holt (1992) have argued that milk production in particular is a likely candidate for actual chaotic dynamics based on a sound theoretical model and empirically realistic parameter estimates. Fourteen year production lags with a nonlinear market supply function due to diminishing returns combined with sufficiently inelastic demand generate apparent SDIC for this market.

Yet another factor complicating empirical tests for chaos is bias arising from aggregation procedures used in data construction, although Barnett and Hinich (1992) argue that this is not a problem for Divisia index measures. This has further tended to push the search for economic chaos towards well-defined microeconomic series such as pork and milk production. Furthermore it has been argued that unstable or chaotic dynamics may be more likely for more localized systems. Thus Dendrinos with Mulalley (1985) argued that unstable dynamics are more common for localized regional systems than for more extended ones. Also Sugihara, Grenfell, and May (1990) found chaotic dynamics for city-by-city measles epidemics but only two-year cycles at the national level.[9]

One way of possibly linking the two approaches to chaotic dynamics may lie in the concept of Kolmogorov entropy (1959) which is given by

$$K = \lim_{r \to 0} \lim_{m \to \infty} \ln [C(r,m)/C(r,m+1)]. \qquad (3.5)$$

This can be interpreted as the gain in information from a finer partition of the space.[10] If SDIC holds then this term is positive because of the Pesin (1977) identity which asserts that K equals the sum of the positive Lyapunov exponents of the process. Brock and Baek (1991) have shown that widely-used estimators of K approach the lower bound of the asymptotic K thus providing possible measure of SDIC derived from dimension estimates.

C. Regime Switching Models

A much noted stylized fact about business cycles is their asymmetry, a characteristic that has also been argued for urban economic cycles although on a much longer time scale. This observation has led to the development of several non linear modelling techniques that seek to predict switching points in a presumed underlying stochastic process.

An influential such approach has been developed by Hamilton (1988). This is a model of a Markov switching process and is given by

$$X_t = n_t + Z_t \tag{3.6}$$

where

$$n_t = \alpha_0 + \alpha_1 s \tag{3.7}$$

and

$$Z_t - Z_{t-1} = \phi_1(Z_{t-1} - Z_{t-2}) + \ldots + \phi_r(Z_{t-r} - Z_{t-r+1}) + \mu_t , \tag{3.8}$$

with the value of s indicating a state or regime of the system. Also s is posited to follow a stochastic process determined by a vector of conditional probabilities for switching from one state to another. Significance of these parameters can be estimated. Using this approach Hamilton (1989) concluded for US real GNP data that the shift from expansion to recession involves a drop in the long-run forecast level of GNP of 3%. Engel and Hamilton (1990) used this approach to argue for the existence of long swings in foreign exchange rates.

A piecewise linear approach developed by Tsay (1989) is the self-exciting threshold autoregressive (SETAR) model.[11] Let $\{X_t\}$ be a stationary time series then a SETAR representation is

$$X_t = \phi(s)_0 + \sum_{i=1}^{p} \phi_i(s) X_{t-i} + e(s)_t, \qquad (3.9)$$

conditional on X_{t-d}, where $e(s)_t$ are white noise processes, $s = 1,\ldots,k$ are regimes, d is the delay parameter, and r is a k-1 vector of threshold parameters. Potter (1991b) has used this approach to model US GNP coming to the conclusion that the post-World War II US economy has been more stable than the pre-war US economy. Ham and Sayers (1992) have shown the superiority of this approach to simple linear models for modeling sectoral unemployment rates in the US.

A generalization of this approach is to allow for a continuous transition from one regime to another. This has been developed by Terasvirta and Anderson (1991) as the smooth transition autoregressive (STAR) model. Equation (3.9) is modified by multiplying the state terms by a transition function which itself becomes an object of testing. For 13 OECD countries they compared logistic and exponential formulations with linear models, obtaining mixed results.

D. Modelling Conditional Variance

Certainly the most widely used non linear time series modelling technique in economics has been to allow for variable variance.[12] A motivation for this in most asset markets is the clear evidence for leptokurtosis, or fat-tailed distributions, consistent with occasional periods of increased volatility. Mandelbrot (1963) argued for the use of the asymptotically infinite variance Pareto-Levy distribution. But the most widely used approach has been to allow for finite but varying variance in the form of autoregressive conditional heteroskedasticity (ARCH) effects. Engle (1982) developed the original ARCH model as

$$x_t = \sigma_t u_t, \qquad (3.10)$$

$$\sigma_t^2 = \phi_0 + \phi x_{t-1}^2 \qquad (3.11)$$

where σ_t is the standard deviation of $\{X\}$ in time t and u_t is an iid standard normal stochastic process.

Bollerslev (1986) expanded this to Generalized ARCH (GARCH) by making σ_t a function of its own past:

$$\sigma_t^2 = \phi_0 + \phi x_{t-1}^2 + \Psi \sigma_{t-1}^2. \tag{3.12}$$

Nelson (1991) developed exponential GARCH (EGARCH) by using the log of σ_t^2:

$$\ln \sigma_t^2 = \phi_0 + \phi \, |x_{t-1}/\sigma_{t-1}| + \Psi \ln \sigma_{t-1}^2 + \gamma \, x_{t-1}/\sigma_{t-1}. \tag{3.13}$$

Hsieh (1991) has argued that this form is the most powerful for explaining non linearities in stock returns of any available, including chaotic dynamics and several we have not (and will not) discuss.[13]

Despite the great success of this category of approaches in analyzing economic time series a question arises in terms of this paper. Most of these series are very high frequency; some of the stock series analyzed by Hsieh (1991) were at 20 second intervals. But time series in spatial dynamic models are generally of much lower frequency and in some cases possibly of very low frequency. This brings us back to the issue of time scales.

Nevertheless it would seem that the ARCH approach might still have some relevance in a dynamic context where evolutionary processes and structural changes are occurring. One theme of such arguments that we shall develop later is that volatility may well increase near major bifurcation points in an evolutionary process. Thus a lower frequency ARCH type approach might be revealing in certain cases.

E. Adaptive Evolutionary Approaches

A recent development has been the appearance within economics of econometric techniques that imply multi-layered feedforward and feedbackward relationships. Such structures can be viewed as involving the possibility of learning and adaptation in a dynamic evolutionary process. The most widely used such technique has been artificial neural networks (ANN) presented in Rummelhart and McClelland (1986) and derived initially from models of parallel processing in brains with numerous applications in many fields.

A purely feedforward architecture for a single output o[14] can be given by an n-vector x of input variables, a function H(x) that activates q hidden units according to weights γ_i giving hidden unit outputs of h_i, and an output activation function G[15] that transmits the hidden unit activations according to weights β_i. Thus:

$$0 = G(\beta_0 + \sum_{i=1}^{q} \beta_i H(\gamma_{i0} + \sum_{j=1}^{n} \gamma_{ij} x_j)). \qquad (3.14)$$

Viewing this as $f(x, \phi)$ where ϕ is the vector of parameters β's and γ's, Hornick, Stinchcombe, and White (1989) have shown that this can approximate a large class of functions arbitrarily well as long as q is suitably large.

Feedbackward or backward propagation, also known as "backprop" (Baum, 1988), leads to recurrent ANN's when combined with feedforward effects. A fairly simple extension of the above is to allow a time delayed feedback from the hidden activations h_i according to weights δ_i. This leads to

$$0_t = G(\beta_0 + \sum_{i=1}^{q} \beta_1 H(\gamma_{i0} + \sum_{j=1}^{n} \gamma_{ij} x_{j,t} + \sum_{l=1}^{q} \delta_{il} h_{l,t-1})). \qquad (3.15)$$

However the final term implies an expansion backwards in time to the beginning of the series of x_t in terms of causation because

$$\sum_{l=1}^{q} \delta_{il} h_{l,t-1} = \sum_{l=1}^{q} \delta_{il} H(\gamma_{10} + \sum_{k=1}^{n} \gamma_{lk} x_{k,t-1} + \sum_{m=1}^{q} \delta_{lm} h_{m,t-2}). \qquad (3.16)$$

This can be summarized as $g(x^t, \tilde{\theta})$ where x^t is the complete history of the input vectors and $\tilde{\theta}$ is the full vector of parameters β, γ, δ.

A criterion for selecting efficient network structure used by econometricians is the Schwarz (1978) Information Criterion which Rissanen (1987) has shown to be asymptotically equivalent to model stochastic complexity. The first effort to use ANN in applied econometrics was White (1988) for IBM stock returns. ANN did not show much power in that test. More recently Kuan and Liu (1992) have shown ANN models beating random walk models in forecasting foreign exchange rates. This is a potentially exciting development insofar as the forecast power of most non linear time series techniques compared to the random walk has not been impressive for that particular time series (Diebold and Nason, 1990; Rosser, 1991, Chap. 15).

3.3   THEORETICAL ISSUES IN NONLINEAR DYNAMICAL ECONOMICS

A.   Adaptive Evolutionary Processes

A standard remark in the econometrics literature regarding artificial neural networks (ANN) has been that it is a "black box" technique. Inputs are transformed into outputs through a multiple layer of hidden structures. This makes it difficult to ascertain the equilibrium, if any, characteristics of the system as well as the qualitative dynamics. Nevertheless the ANN approach appears to be broadly consistent with the rapidly developing school of thought of evolutionary economics.[16]

If one is considering an evolving spatial economic system with complex feedforward and backprop relationships, then one must contemplate the overall interaction and self-organization of all the components of the system. This becomes more serious in a non linear system characterized by increasing returns to scale both internally and externally because of the existence of multiple equilibria, path dependence, and the possibility of "lock-in" to a sub-optimal path (Arthur, 1988).[17] The length of the time delays in the backprop process and the degree of interconnectedness within the system can become crucially important elements in the dynamic process.[18]

From an evolutionary perspective this feedback adaptation process can be viewed in terms of the complexity of the sytem as defined by the number of elements N and the degree of dependence of an element's fitness on other elements given by the number of other elements its fitness depends on K (Kauffman, 1988). Two extreme cases are K = 0 and K = N - 1. In the first each part is independent of the rest. This leads to a "correlated fitness landscape" because changing one part changes overall fitness by no more than 1/N. In the second each part depends on all the others. This leads to an "uncorrelated fitness landscape" because changing one part alters the contributions of all parts in a way essentially random relative to the original situation.[19]

Kauffman (1988) shows that the mean time it takes for a system to reach a local optimum assuming random interactions tends to increase with N and decrease with K. As N and K increase the fitness of local optima diverge from the global optimum and approach the mean of the space. For the completely uncorrelated case time to reach an improved variant doubles after each improvement (Kauffman and Levin, 1987). Thus the expected number of improvement steps S as a function of the number of "tries" G is

$$S = \log_2(G). \tag{3.17}$$

This approach can be made analogous to the ANN approach in the following way. Fitness can be viewed as equivalent to output, N as equivalent to the xt input vector, and K as equivalent to the parameter vector. Thus the more inputs the slower the adaptiveness of the system and the greater the feedforward and feedbackward complexity the more rapid the adaptiveness. On the other hand greater complexity increases the divergence between local and global optima.

This parallels the issue of sub-optimality in the path dependence literature. Greater complexity of interrelationships may lead to more rapid adjustment to a local optimum, but it will be more likely to be not only not the global optimum but farther away from it as well. There is every reason to expect this argument to carry over to dynamically evolving spatial economic systems also.[20]

B.   Self-Organized Criticality

The concept of a self-organizing system is deeply rooted in economic thought. Although he never used the term, Adam Smith's (1776) view of economic dynamics can be claimed to be consistent with such an idea. Although he did not present a mathematical model, Hayek (1948) used the term in his Austrian defense of laissez-faire as leading to evolutionary self-organization, an argument updated by Lavoie (1989) in the context of more modern theories of chaos and out-of-equilibrium dynamics.

The modern more formal formulation of self-organization theories has developed along two closely related strands. One is the out-of-equilibrium phase transition approach developed by Nicolis and Prigogine (1977). This has had great influence in spatial economics through the "Brussels School" (Allen and Sanglier, 1981), although less so in other areas of economics.[21] The other is the synergetics approach developed by Haken (1977). This was initially used to analyze technological innovation and diffusion questions (Silverberg, 1984; Silverberg, Dosi, and Orsenigo, 1988), but has since been argued to be a possible unifying methodology for the analysis of nonlinear evolutionary economic dynamics in general (Zhang, 1991).

It can be argued that both entail the idea of a possibly heightened volatility during major systemic structural change. For the former this is associated with random fluctuations near a phase transition point if the system is jumping back and forth across the point. For the latter this is associated with chaotic dynamics arising from the "revolt of the slaved variables" as order parameters become endogenous variables temporarily. In both cases the ARCH approach discussed above might be an appropriate empirical technique for identification of such phenomena.[22]

On the other hand as argued by members of this workshop during the the presentation of the first draft of this paper, these approaches may not imply such an outcome. There may be no change in variance or even a decrease in variance as bifurcation point is approached. The latter idea corresponds with the argument of Holling (1973) regarding terrestial ecosystems that there is a tradeoff between stability and resiliency. A system that becomes extremely stable, such as a climax forest (or the former Soviet economy), is more in danger of a major collapse or structural change than one that is fluctuating more. More recently Holling (1987, 1992) has argued that there are four stages in ecosystem evolution during two of which major structural changes can occur. In one rigidity precedes reorganization, but in the other reorganization is associated with high variance and disorganization. Thus either a decrease or an increase in variance could precede major structural change.

A more recent effort to model self-organization in economic dynamics depends on the notion of self-organized criticality, originally developed to explain sandpiles and earthquakes (Bak and Chen, 1991) and drawing heavily on the Pareto-Levy distribution advocated by Mandelbrot (1963). Bak, Chen, Scheinkman, and Woodford (1992) present a model of multi-layered stages of production displayable on an L by L lattice with N buyers making random purchases. A firm on the ith row and jth column will have $x_{i,j}(t)$

inventory at time t that can be either 0 or 1 with y similarly indexed being its production and Y being aggregate production. Firms face sharply non-convex cost functions due to indivisibilities. This will lead to periodic "avalanches" as aggregate production makes large responses to small final demand fluctuations at critical values.[23]

In contrast to other approaches to self-organization there is a probability distribution of these critical values of varying degrees of intensity. They investigate these distributions under various assumptions. Letting τ be a parameter dependent on the dimensionality of the lattice they obtain the following asymptotic distribution for y given a large N:

$$Q(y) = 1/N^{1/\tau} \, G(y/N^{1/\tau}). \tag{3.18}$$

G is a stable Paretian distribution whose asymptotic behavior for large $y/N^{1/\tau} = x$ is given by

$$G(x) = \sin(\pi\tau/2) \, \Gamma \, (1+\tau) / x^{1+\tau}, \tag{3.19}$$

with Γ being the gamma function of probability theory. This distribution is skewed and will for the case of N shocks and L production stages possess a median of $L^{1/2}$.

This approach is very consonant with the out-of-equilibrium phase transition view. Generally the system will not be in a global optimum and indeed will generally be moving away from that locally most of the time. The metaphor is the sand pile. The lowest energy state is being completely flat. But as sand is dropped on the pile it tends to accumulate upward, increasing its slope until an avalanche occurs, which then lowers it somewhat, but not all the way to zero or long-run equilibrium flatness.

This approach has been explicitly conjoined to the adaptive evolutionary approach. Langton (1990) and Kauffman and Johnsen (1991) have shown that evolution may be most likely to occur at "the edge of chaos" in a poised state that resembles self-organized criticality. Kauffman and Johnsen identify the associated avalanches with waves of extinctions in rapid evolutionary contexts.

C. Chaotic Hysteresis

Questions of how self-organization and evolutionary feedback in nonlinear economic dynamics lead to larger scale structural economic changes can be explained by a concept first labeled by this author (1991, Chap. 17) as "chaotic hysteresis." It was initially viewed as a way of integrating a catastrophe theory view of large scale morphogenesis with the emergence of chaotic dynamics at or near the bifurcation points. In this respect the concept is very close in spirit to both the Brussels School and synergetics approaches.

The simplest version of such a model is depicted in Figure 1 and reflects a cyclical model of y as a function of x. The cycle involves a hysteresis effect with associated catastrophic leaps at critical points. For spatial models such a hysteresis system has been posed without the chaotic dynamics in the "logistical networks" model of Andersson (1986) where x is scale of logistical networks infrastructure and y is population or possibly the level of economic activity. In the chaotic hysteresis version chaotic dynamics emerge in the neighborhoods of critical points.

Although he did not label it as chaotic hysteresis, Puu (1990) developed a nonlinear multiplier-accelerator model that exhibits behavior consistent with such a view. The model is given by

$$Y_t = C_t + I_t, \qquad (3.20)$$

$$C_t = (1-s)Y_{t-1}, \qquad (3.21)$$

$$I_t = v(I_{t-1} - I_{t-2}) - v(I_{t-1} - I_{t-2})^{3}.^{25} \qquad (3.22)$$

Letting $z_t = I_t - I_{t-1}$ the system reduces to

$$I_t = I_{t-1} + z_t, \qquad (3.23)$$

$$z_t = v(z_{t-1} - z_{t-1}^3) - sI_{-1}. \qquad (3.24)$$

At v = 2 and s vanishingly small Puu (1990) shows a pattern in (I, z) space that looks

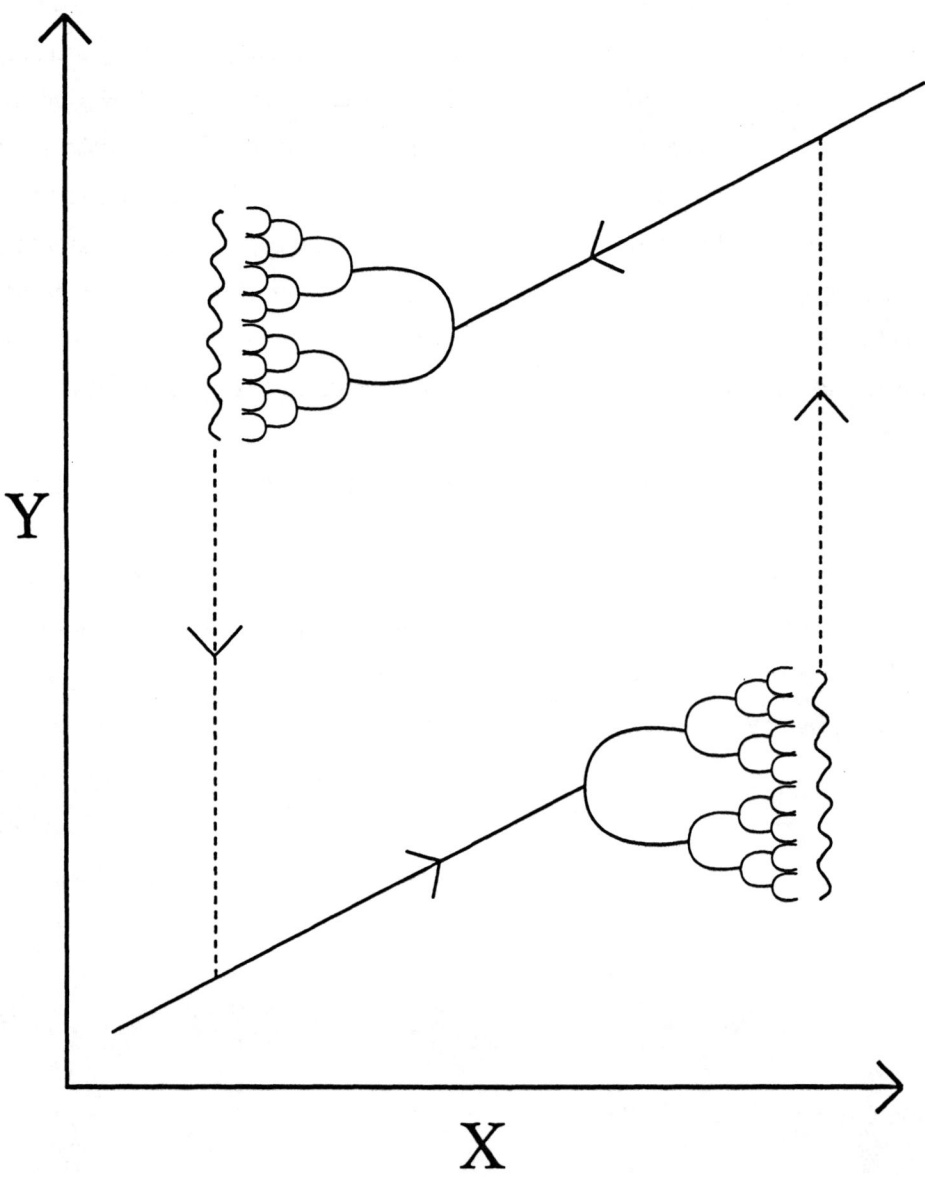

Figure 1   Chaotic Hysteresis

much like Figure 1. This model exhibits chaotic dynamics occuring <u>after</u> the discontinuous leap of z with a period-halving transition out of chaotic dynamics.

Rosser and Rosser (1992a, b) adapted this model to a model of capital self-ordering of Forrester (1977) and Sterman (1985) to explain long investment wave cycles in command socialist economies. Such long investment cycles may be associated with the kind of large-scale infrastructure investments implicit in the Andersson (1986) spatial model of logistical networks. Rosser and Rosser have argued that systemic crisis and the possibility of systemic collapse and transformation is greatest in the zone of chaotic dynamics. Such an argument is clearly in the spirit of the self-organization and synergetics views of economic evolution.[26]

3.4   CONCLUSION

This paper has surveyed a variety of techniques and concepts used in the modelling of nonlinear economic dynamical systems. Among econometric techniques have been those looking at chaotic dynamics, structural breaks in the form of regime switching models, ARCH and related effects, and feedforward and backward systems in the form of artificial neural networks. Many of these are potentially usable for the analysis of spatial economic systems experiencing nonlinear evolutionary dynamics involving structural breaks or highly volatile self-organization processes with adaptive mechanisms.

The theory of adaptive processes was related to the ANN framework as was a recent model of self-organized criticality and a model of chaotic hysteresis. All of these would seem to have applications in the spatial evolutionary context.

The final discussion of chaotic hysteresis argued that such processes may be operating over long wave time scales, especially in a logistical networks infrastructure investment context. However at the other extreme we see that unstable or chaotic dynamics may be most likely to operate at a highly localized and high frequency ranges. Thus there might be a synthesis in the form of a hierarchy of cycles associated with different time scales of evolutionary dynamics. Such an approach is consistent with Schumpeter's (1939) vision of multiple hierarchical business cycles (Rosser and Rosser, 1992c). Such hierarchies of time scales have been modeled explicitly for spatial systems by Wegener (1982), Wegener, Gnad, and Vanvahme (1986), and Batten and Johannson (1987), the latter explicitly within the context of synergetics.[27]

REFERENCES

Allen, Peter M. and M. Sanglier, 1981, Urban Evolution, Self Organization, and Decision Making, **Environment and Planning A** 13, pp. 167-183.

Andersson, Ake E., 1986, The Four Logistical Revolutions, **Papers of the Regional Science Association** 59, pp. 1-12.

Arthur, W.B., 1988, Self-Reinforcing Mechanisms in Economics, in **The Economy as an Evolving Complex System**, P.W. Anderson, K.J. Arrow, and D. Pines, (eds.), Santa Fe Institute Series, Addison-Wesley, Redwood City, pp. 9-31.

Artstein, Z., 1983, Irregular Cobweb Dynamics, **Economic Letters** 11, pp. 15-17.

Bak, P. and Kan Chen, 1991, Self Organized Criticality, **Scientific American** 264, p.46.

Bak, P., Kan Chen, Jose A. Scheinkman, and M. Woodford, 1992, Self Organized Criticality and Fluctuations in Economics, mimeo, Santa Fe Institute.

Barnett, W.A., A. Gallant, M.J. Hinich, and M.J. Jensen, 1992, Robustness of Nonlinearity and Chaos Tests to Measurement Error, Inference Method, and Sample Size, mimeo, Washington University, North Carolina State University, and University of Texas.

Barnett, W.A., A. Gallant and M.J. Hinich, 1992, Has Chaos Been Discovered with Economic Data? in **Evolutionary Dynamics and Nonlinear Economics**, R.H. Day and P. Chen, (eds.), Oxford University Press, Oxford.

Batten, D. and B. Johannson, 1987, Dynamics of Metropolitan Change, **Geographical Analysis** 19, pp. 189-199.

Baum, E.B., 1988, Neural Nets for Economists, in **The Economy as an Evolving Complex System**, P.W. Anderson, K.J. Arrow, and D. Pines, (eds.), Santa Fe Institute Series, Addison-Wesley, Redwood City, pp. 33-48.

Baumol, W.J. and J. Benhabib, 1989, Chaos: Significance, Mechanism, and Economic Applications, **Journal of Economic Perspectives** 3, pp. 77-105.

Benhabib, J. and R.H. Day, 1980. Erratic Accumulation, **Economic Letters** 6, pp. 113-117.

Boldrin, M. and M. Woodford, 1990, Equilibrium Models Displaying Endogenous Fluctuations and Chaos: A Survey, **Journal of Monetary Economics** 25, pp. 189-222.

Bollerslev, T., 1986, Generalized Autoregressive Conditional Heteroskedasticity, **Journal of Econometrics** 31, pp. 307-327.

Bollerslev, T., R.Y. Chou, N. Jayaraman, and K.F. Kroner, 1990, ARCH Modeling in Finance: A Selective Review of the Theory and the Empirical Evidence, with Suggestions for Future Research, mimeo, University of Arizona.

Brock, W.A., 1991, Understanding Macroeconomic Time Series Using Complex Systems Theory, **Structural Change and Economic Dynamics** 2, pp. 119-141.

Brock, W.A. and Ehung G. Baek, 1991, Some Theory of Statistical Inference for Nonlinear Science, **Review of Economic Studies** 58, pp. 697-716.

Brock, W.A., W.D. Dechert, and Jose A. Scheinkman, 1987, A Test for Independence Based on the Correlation Dimension, **SSRI Paper No. 8702**, University of Wisconsin-Madison.

Brock, W.A., D. Hsieh, and B. LeBaron, 1991, **Nonlinear Dynamics, Chaos, and Instability**, MIT Press, Cambridge, MA.

Brock, W.A. and S.M. Potter, 1991, Diagnostic Testing for Nonlinearity, Chaos, and General Dependence in Time Series Data, in **Proceedings of the 1990 Workshop on Nonlinear Modeling and Forecasting**, S. Eubank and M. Casdagli, (eds.), Santa Fe Institute Series, Addison-Wesley, Redwood City.

Brock, W.A. and C.L. Sayers, 1988, Is the Business Cycle Characterized by Deterministic Chaos? **Journal of Monetary Economics** 22, pp. 71-79.

Casdagli, M., S. Eubank, J. Doyne Farmer, and J. Gibson, 1991, State Space Reconstruction in the Presence of Noise, **Physica D** 51, pp. 52-98.

Chavas, J.-P. and M.T. Holt, 1991, On Nonlinear Dynamics: The Case of the Pork Cycle, **American Journal of Agricultural Economics** 73, pp. 819-828.

Chavas, J.-P. and M.T. Holt, 1992, Market Instability and Nonlinear Dynamics, **American Journal of Agricultural Economics**, forthcoming.

Chiarella, C., 1988, The Cobweb Model: Its Instability and the Onset of Chaos, **Economic Modelling** 5, pp. 377-384.

Cleveland, W.A., 1979, Robust Locally Weighted Regression and Smoothing Scatterplots, **Journal of the American Statistical Association** 74, pp. 829-836.

Day, R.H. and G. Piangiani, 1991, Statistical Dynamics and Economics, **Journal of Economic Behavior and Organization** 16, pp. 37-83.

Dendrinos, D.S. with Henry Mulally, 1985, **Urban Evolution: Studies in the Mathematical Ecology of Cities**, Oxford University Press, Oxford.

Diebold, F.X. and J.M. Nason, 1990, Nonparametric Exchange Rate Prediction, **Journal of International Economics** 28, pp. 315-332.

Durlauf, S.N., 1991, Nonergodic Economic Growth, **Technical Report No. 7**, Stanford Institute for Theoretical Economics.

Eckmann, J.-P., S. Ollifson Kamphorst, D. Ruelle, and S. Ciliberto, 1986, Lyapunov Exponents from a Time Series, **Physical Review A** 34, pp. 4971-4979.

Eckmann, J.-P., S. Ollifson Kamphorst, D. Ruelle, and J.A. Scheinkman, 1988, Lyapunov Exponents for Stock Returns, in **The Economy as an Evolving Complex System**, P.W. Anderson, K.J. Arrow, and D. Pines, (eds.), Santa Fe Institute Series, Addsion-Wesley, Redwood City, pp. 301-304.

Eckmann, J.-P. and D. Ruelle, 1985, Ergodic Theory of Chaos and Strange Attractors, **Review of Modern Physics** 57, pp. 617-656.

Engel, C. and J.D. Hamilton, 1990, Long Swings in the Exchange Rate: Are they in the Data and Do Markets Know it? **American Economic Review** 80, pp. 689-713.

Engle, R.F., 1982, Autoregressive Conditional Heteroskedasticity with Estimates of the Variance of United Kingdom Inflation, **Econometrica** 50, pp. 987-1007.

Engle, R.F. and T. Bollerslev, 1986, Modeling the Persistence of Conditional Variances, **Econometric Reviews** 5, pp. 1-50.

Engle, R.F., D.M. Lillien, and R.P. Robins, 1987, Estimating Time Varying Risk Premia in the Term Structure: The ARCH-M Model, **Econometrica** 55, pp. 391-407.

Fischer, E.O. and W. Jammernegg, 1986, Empirical Investigation of a Catastrophe Theory Extension of the Phillips Curve, **Review of Economics and Statistics** 68, pp. 9-17.

Forrester, J.W., 1977, Growth Cycles, **De Economist** 125, pp. 525-543.

Frank, M.Z. and T. Stengos, 1989, Measuring the Strangeness of Gold and Siver Rates of Return, **Review of Economic Studies** 56, pp. 553-568.

Grandmont, J.-M., 1985, On Endogenous Competitive Business Cycles, **Econometrica** 53, pp. 995-1045.

Grassberger, P. and I. Procaccia, 1983, Measuring the Strangeness of Strange Attractors, **Physica 9D**, 189-208.

Gregory-Allen, R.B. and G.V. Henderson, Jr., 1991, A Brief Review of Catastrophe Theory and a Test in a Corporate Failure Context, **The Financial Review** 26, pp. 127-155.

Guckenheimer, J. and P. Holmes, 1983, **Nonlinear Oscillations, Dynamical Systems, and Bifurcations of Vector Fields**, Springer-Verlag, Berlin.

Haken, H., 1977, **Synergetics**, Springer-Verlag, Berlin.

Ham, M.R. and C.L. Sayers, 1992, Testing for Nonlinearities in United States Unemployment by Sector, mimeo, University of Virginia and University of Houston.

Hamilton, J.D., 1988, Rational-Expectations Econometric Analysis of Changes in Regime: An Investigation of the Term Structure of Interest Rates, **Journal of Economic Dynamics and Control** 12, pp. 385-423.

Hamilton, J.D., 1989, A New Approach to the Economic Analysis of Nonstationary Time Series and the Business Cycle, **Econometrica** 57, pp. 357-384.

Hayek, F.A., 1948, **Individualism and Economic Order**, University of Chicago Press, Chicago.

Higgins, M.L. and A.K. Bera, 1992, A Class of Nonlinear ARCH Models, **International Economic Review** 33, pp. 137-158.

Holling, C.S., 1973, Resilience and Stability of Ecological Systems, **Annual Review of Ecology and Systematics** 4, pp. 1-23.

Holling, C.S., 1987, Simplifying the Complex: The Paradigms of Ecological Function and Structure, **European Journal of Operations Research** 30, pp. 139-146.

Holling, C.S., 1992, Cross-Scale Morphology, Geometry and Dynamics of Ecosystems, **Ecological Monographs**, forthcoming.

Hornik, K., M. Stinchcombe, and H. White, 1989, Multi-Layer Feedforward Networks are Universal Approximators, **Neural Networks** 2, pp. 359-366.

Hsieh, D.A., 1989, Testing for Nonlinear Dependence in Daily Foreign Exchange Rates, **Journal of Business** 62, pp. 339-368.

Hsieh, D.A., 1991, Chaos and Nonlinear Dynamics: Application to Financial Markets,

Journal of Finance 46, pp. 1839-1877.

Invernizzi, S. and A. Medio, 1991, On Lags and Chaos in Economic Dynamic Models, **Journal of Mathematical Economics** 20, pp. 521-550.

Jensen, R.V. and R. Urban, 1984, Chaotic Price Behavior in a Non-Linear Cobweb Model, **Economic Letters** 15, pp. 235-240.

Kauffman, S.A., 1988, The Evolution of Economic Webs, in **The Economy as an Evolving Complex System**, P.W. Anderson, K.J. Arrow, and D. Pines, (eds.), Santa Fe Institute Series, Addison-Wesley, Redwood City, pp. 125-146.

Kauffman, S.A. and S. Levin, 1987, Towards a General Theory of Adaptive Walks on Rugged Landscapes, **Journal of Theoretical Biology** 128, pp. 11-45.

Kauffman, S.A. and S. Johnsen, Coevolution to the Edge of Chaos: Coupled Fitness Landscapes, Poised States, and Coevolutionary Avalanches, **Journal of Theoretical Biology** 149, pp. 467-505.

Kelsey, D., 1988, The Economics of Chaos or the Chaos of Economics, **Oxford Economic Papers** 40, pp. 1-31.

Kolmogorov, A.N., 1959, Epsilon-entropy and Epsilon-capacity of Sets in Functional Spaces, **Uspekhi Matematicheskikh Nauk** 14, pp. 3-86.

Kuan, Chung-Ming and Tung Liu, 1992, Forecasting Exchange Rates Using Feedforward and Recurrent Neural Networks, **Faculty Working Paper 92-0128**, Bureau of Economic and Business Research, University of Illinois.

Langton, C.G., 1990, Computation at the Edge of Chaos: Phase Transitions and Emergent Computation, **Physica D** 42, pp. 12-37.

Lavoie, D., 1989, Economic Chaos or Spontaneous Order? Implications for Political Economy of the New View of Science, **Cato Journal** 8, pp. 613-635.

LeBaron, B., 1991, Empirical Evidence for Nonlinearities and Chaos in Economic Time Series: A Summary of Recent Results, **SSRI Paper No. 9117**, University of Wisconsin-Madison.

Lorenz, H.-W., 1989, **Nonlinear Dynamical Economics and Chaotic Motion**, Springer-Verlag, Berlin.

Mandelbrot, B.B., 1963, The Variation of Certain Speculative Prices, **Journal of Business** 36, pp. 394-419.

May, R.M., 1974, Biological Populations with Non-Overlapping Generations: Stable Points, Stable Cycles, and Chaos, **Science** 186, pp. 645-647.

May, R.M., 1976, Simple Mathematical Models with Very Complicated Dynamics, **Nature** 261, 459-467.

Mayfield, E.S. and B. Mizrach, 1989, On Determining the Dimension of Real-Time Stock Price Data, Department of Economics Working Paper No. 187, Boston College.

Mittnik, S., 1992, Nonlinear Output and Unemployment Dynamics in U.S. Business Cycles, mimeo, SUNY-Stony Brook.

Nelson, C. and C. Plosser, 1982, Trends and Random Walks in Macroeconomic Time Series: Some Evidence and Implications, **Journal of Monetary Economics** 10, pp. 139-162.

Nelson, D.B., 1991, Conditional Heteroskedasticity in Asset Returns: A New Approach, **Econometrica** 59, pp. 347-370.

Nicolis, G. and I. Prigogine, 1977, **Self-Organization in Non-Equilibrium Systems: From Dissipative Structures to Order Through Fluctuations**, Wiley-Interscience, New York.

Nijkamp, P. and A. Reggiani, 1992, Spatial Competition and Ecologically Based Socio-Economic Models, **Socio Spatial Dynamics**, vol. 3, no. 2, pp. 89-109.

Nychka, D., S. Ellner, R. Gallant, and D. McCaffrey, 1992, Finding Chaos in Noisy Systems, **Journal of the Royal Statistical Society B** 54, pp. 399-426.

O'Neill, R.V., A.R. Johnson, and A.W. King, 1989, A Hierarchical Framework for the Analysis of Scale, **Landscape Ecology** 3, pp. 193-205.

Pesin, Ya.B., 1977, Characteristic Lyapunov Exponents and Smooth Ergodic Theory, **Russian Mathematical Surveys** 32, pp. 55-114.

Potter, S.M., 1991a, Nonlinear Impulse Response Functions, mimeo, University of California-Los Angeles.

Potter, S.M., 1991b, A Nonlinear Approach to U.S. GNP, mimeo, University of California-Los Angeles.

Priestley, M.B., 1988, **Non-Linear and Non-Stationary Time Series Analysis**, Academic Press, New York.

Puu, T., 1990, A Chaotic Model of the Business Cycle, **Occasional Paper Series on Socio-Spatial Dynamics** 1, pp. 1-19.

Puu, T., 1991, Chaos in Business Cycles, Chaos, **Solitons and Fractals** 1, pp. 457-473.

Rand, D., 1978, Exotic Phenomena in Games and Duopoly Models, **Journal of Mathematical Economics** 5, pp. 139-154.

Rissanen, J., 1987, Stochastic Complexity (with discussions), **Journal of the Royal Statistical Society B**, 49, pp. 223-239 and pp. 252-265.

Romer, P., 1986, Increasing Returns and Long Run Growth, **Journal of Political Economy** 94, pp. 1002-1037.

Rosser, J.B., Jr., 1990, Approaches to the Analysis of the Morphogenesis of Regional Systems, **Occasional Paper Series on Socio-Spatial Dynamics** 1, pp. 75-102.

Rosser, J.B., Jr., 1991, **From Catastrophe to Chaos: A General Theory of Economic Discontinuities**, Kluwer Academic Publishers, Boston.

Rosser, J.B., Jr., 1992a, Chaos Theory and Rational Expectations in Economics, in **Chaos Theory in the Social Sciences**, E. Elliott and L.D. Kiel, (eds.), University of Michigan Press, Ann Arbor, forthcoming.

Rosser, J.B., Jr., 1992b, The Dialogue Between the Economic and the Ecologic Theories of Evolution, **Journal of Economic Behavior and Organization** 17, pp. 195-215.

Rosser, J.B., Jr., 1992c, Morphogenesis of Urban Historical Forms, **Socio-Spatial Dynamics** 3, pp. 17-34.

Rosser, J.B., Jr. and M. Vcherashnaya Rosser, 1992a, Long Wave Chaos and Systemic Economic Transformation, mimeo, James Madison University.

Rosser, J.B., Jr. and M. Vcherashnaya Rosser, 1992b, Chaotic Hysteresis and Systemic Economic Transformation, mimeo, James Madison University.

Rosser, J.B., Jr. and M. Vcherashnaya Rosser, 1992c, Schumpeterian Cycles and the Transition from Socialism to Capitalism, mimeo, James Madison University.

Rummelhart, D.E. and J.L. McClelland, (eds.), 1986, **Parallel Distributed Processing**, vol. 1, MIT Press, Cambridge, MA.

Sayers, C.L., 1988, Work Stoppages: Exploring the Nonlinear Dynamics, mimeo, University of Houston.

Scheinkman, J.A. and B. LeBaron, 1989, Nonlinear Dynamics and Stock Returns, **Journal of Business** 62, pp. 311-337.

Schumpeter, J.A., 1939, **Business Cycles**, McGraw-Hill, New York.

Schwarz, G., 1978, Estimating the Dimension of a Model, **Annals of Statistics** 6, pp. 461-464.

Sherrington, D. and S. Kirkpatrick, 1975, Solvable Model of a Spin Glass, **Physical Review Letters** 35, pp. 1792.

Silverberg, G., 1984, Embodied Technical Progress in a Dynamic Model: The Self-Organization Paradigm, in **Nonlinear Models of Fluctuating Growth**, R.M. Goodwin, M. Kruger, and A. Vercelli, (eds.), Springer-Verlag, Berlin, pp. 192-208.

Silverberg, G., G. Dosi, and L. Orsenigo, 1988, Innovation, Diversity and Diffusion, **Economic Journal** 98, pp. 1032-1054.

Smith, A., 1776, **An Inquiry into the Nature and Causes of the Wealth of Nations**, Strahan and Cadell, London.

Sterman, J.D., 1985, A Behavioral Model of the Economic Long Wave, **Journal of Economic Behavior and Organization** 6, pp. 17-53.

Stutzer, M.J., 1980, Chaotic Dynamics and Bifurcations in a Macro Model, **Journal of Economic Dynamics and Control** 2, pp. 353-376.

Sugihara, G., B. Grenfell, and R. May, 1990, Distinguishing Error from Chaos in Ecological Time Series, mimeo, University of California-San Diego, Cambridge University, and Oxford University.

Terasvirta, T. and H.M. Anderson, 1991, Modelling Nonlinearities in Business Cycles Using Smooth Transition Autoregressive Models, mimeo, Research Institute of the Finnish Economy and University of California-San Diego.

Tong, H. and K.S. Lim, 1980, Threshold Autoregression, Limit Cycles and Cyclical Data, **Journal of the Royal Statistical Society** B 42, pp. 245-292.

Tsay, R.S., 1989, Testing and Modeling Threshold Autoregressive Processes, **Journal of the American Statistical Association** 84, pp. 231-240.

Wales, D.J., 1991, Calculating the Rate of Loss of Information from Chaotic Time Series by Forecasting, **Nature** 350, pp. 485-488.

Wegener, M., 1982, Modeling Urban Decline: A Multi-Level Economic-Demographic Model of the Dortmund Region, **International Regional Science Review** 7, pp. 21-41.

Wegener, M., F. Gnad, and M. Vannahme, 1986, The Time Scale of Urban Change, in **Advances in Urban Systems Modelling**, B. Hutchinson and M. Batty, (eds.), North-Holland, Amsterdam, pp. 175-198.

Weidlich, W. and G. Haag, 1983, **Concepts and Models of a Quantitative Sociology, The Dynamics of Interaction Populations**, Springer-Verlag, Berlin.

White, H., 1988, Economic Prediction Using Neural Networks: The Case of IBM Stock Prices, in **Proceedings of the IEEE Second International Conference on Neural Networks**, II, SOS Printing, San Diego, pp. 451-458.

Wold, H.O.A., 1938, **A Study in the Analysis of Stationary Time Series**, Almquist and Wiksell, Uppsala.

Wolf, A.J., J. Swift, H. Swinney, and J. Vastano, 1985, Determining Lyapunov Exponents from a Time Series, **Physica D** 16, pp. 285-317.

Zhang, Wei-Bin, 1991, **Synergetic Economics: Time and Change in Nonlinear Economics**, Springer-Verlag, Berlin.

Zhang, Wei-Bin, 1992, A Development Model of Developing Economies with Capital and Knowledge Accumulation, **Journal of Economics** (Zeitschrift fur Nationalökonomie) 55, pp. 43-63.

NOTES

1. The author is Professor of Economics at James Madison University in Harrisonburg, Virginia, USA. He wishes to acknowledge receipt of useful materials or comments from William Barnett, William Brock, Jean-Paul Chavas, Rod Cross, Richard Day, Dimitrios Dendrinos, Ronald Gallant, James Hamilton, C.S. Holling, Matthew Holt, Chung-Ming Kuan, Blake LeBaron, Peter Nijkamp, Simon Potter, Tonu Puu, Aura Reggiani, Chera Sayers, Halbert White, Mark White, and the participants in Workshop on Non Linear Evolution of Spatial Economic Systems.

2. This means that all trend and seasonality components have been removed and that no structural changes occur. As the long debate ensuing on Nelson and Plosser (1982) has shown this is a non-trivial issue, but we shall not develop it further here.

3. Although not necessary it is frequently assumed that variance is finite. Allowing asymptotically infinite variance moves one towards Pareto-Levy distributions emphasized by Mandelbrot (1963) and more recently in the self-organizing criticality literature (Bak, Chen, Scheinkman, and Woodford, 1992).

4. This expansion is derived from the Taylor series. Mittnik (1992) has argued that it is a very general non linear form from which most non linear time series models can be derived as special cases.

5. May (1976) first suggested the possibility of such applications in economics. Rand (1978) presented the first microeconomic model and Stutzer (1980) and Benhabib and Day (1980) presented the first macroeconomic models. Interest was especially heightened by the model of Grandmont (1985). The literature on these is now voluminous. Reviews can be found in Kelsey (1988), Baumol and Benhabib (1989), Lorenz (1989), Boldrin and

Woodford (1990), and Rosser (1991).

6. More recently methods using neural nets to estimate Jacobian determinants have been used for calculating Lyapunov exponents from time series (Nychka, Ellner, Gallant, and McCaffrey, 1992). Barnett, Gallant, Hinich, and Jensen (1992) suggest that bootstrapping techniques might be used to generate standard errors for Lyapunov exponent estimates, although the properties of such estimates are unknown in the presence of significant noise.

7. For a thorough discussion of the characteristics of this statistic as well as a summary of applications of it to many economic time series see Brock, Hsieh, and LeBaron (1991).

8. A favourite method has been the locally weighted regression method due initially to Cleveland (1979).

9. See Rosser (1992a) for further discussion.

10. Wales (1991) argues that it also measures the rate of loss of forecast accuracy. Casdagli, Eubank, Farmer, and Gibson (1991) show how forecast accuracy deteriorates in the presence of noise when dimensionality and Lyapunov exponents are both varied.

11. An earlier version of this is the TAR, initially developed by Tong and Lim (1980).

12. The literature on this topic is also voluminous. Reviews include Engle and Bollerslev (1986) and Bollerslev, Chou, Jayaraman, and Kroner (1990).

13. More generalized non linear ARCH forms have been developed by Higgins and Bera (1992). Another approach is to allow for ARCH effects in the mean as well in the variance, the ARCH-M model (Engle, Lillien, and Robins, 1987).

14. This can be generalized to a vector of outputs, but so far all applications in economics have been to single variable output models.

15. A frequently used form for G in applied econometrics is the logistic function. H is often just the identity function.

16. For a discussion of links between current approaches to evolutionary economics and ecological theory see Rosser (1992b).

17. A close relative of the path dependence school is the endogenous growth school (Romer, 1986). It also emphasizes internal and external economies of scale and multiple equilibria but with less of an emphasis on the evolutionary aspect of the non linear dynamic process. See Rosser (1992c) for further discussion.

18. Durlauf (1991) has modeled evolutionary dynamics depending on firms effecting neighboring firms in a non linear feedback framework.

19. This kind of model can also be approached by use of spin glass methods of self-organization in alloys (Sherrington and Kirkpatrick, 1975). Brock (1991) has suggested the application of such models to the study of financial markets.

20. Nijkamp and Reggiani (1992) have shown a spatial evolutionary adaptive model with a chain of niches.

21. A closely related approach emphasizing stochastic aspects more is due to Weidlich and Haag (1983).

22. Such arguments have also been used in the literature on evolutionary hierarchy theory (O'Neill, Johnson, and King, 1989).

23. Clearly there is a certain parallel to the ANN approach here.

24. This is actually an approximation, there being no closed analytical expression for G.

25. This cubic formulation is justified by Puu (1991) as possibly arising from countercylical government investment.

26. Zhang (1992) has developed a cusp catastrophe model of discontinuous changes in knowledge accumulation in China as a function of continuous changes in reform orientation and the degree of openness of the economy. There have been few successful efforts at empirical modelling catastrophes in economics (see Rosser, 1991, Chap. 2), but two such may be Fischer and Jammernegg (1986) and Gregory-Allen and Henderson (1991).

27. See Rosser (1990) for further discussion.

CHAPTER 4
TOWARDS A DYNAMIC DISEQUILIBRIUM THEORY OF ECONOMY
G. Haag, M. Hilliges and K. Teichmann

4.1 INTRODUCTION

Recently the field of disequilibrium economics is becoming more and more attractive. Of course, a theory of equilibrium can be seen as a first approximation of a dynamic theory, because it asserts that a system out of equilibrium must change. The mistake that is often made by practitioners of the comparative static method is to assume that changes out of equilibrium will lead inevitably, with the passage of sufficient time, to equilibrium (R.Day, 1984).

It is possible that a static theory will provide an acceptable prediction of the initial direction of change. It will certainly not be a good theory on the long run (e.g. see the predictions for the evolution of northern Sweden, southern Italy, Berlin).

The path of an economic system can be viewed as a sequence of states in response to the actions taken by the economic agents involved. Those agents or different groups of agents respond to their changing environment, having different choice strategies in mind and taking into account their currently perceived constraints and objectives.

The interrelated decisions in such a complex system constitute the basis of the dynamic evolution of the economy (Haag, 1990). Understanding of dynamic adjustment processes in the context of individual decisions is therefore of crucial importance (Samuelson, 1948, Fischer, 1983, Batten et al, 1985, Barentsen and Nijkamp, 1993). Moreover, it is reasonable to assume that the actions of the agents, based only in part on rational plans (e.g. profit optimization, utility maximization) influence the system through a feedback structure. In general the plans of the decision makers are imperfectly coordinated, based on expectations which are seldom fulfilled, and are not consistent and optimal, at least on the long run. Therefore, an equilibrium situation where demand equals supply and in which technical and social efficiency prevails can be seen as an exceptional event.

Uncertainties in the choice process of individuals require a stochastic treatment of the socio-economic system and its subsystems because the decision makers have imperfect information, limited foresight and changing preferences (Weidlich and Haag, 1983).

Systems composed of subsystems with those inherent nonlinear interactions exhibit in general a variety of instabilities under certain parameter conditions. Phase transitions are regular events if the control parameters of a system pass so called critical values. As a consequence, the evolution of economic systems are in general path dependent.

Events happening in the 'environment' influence the system in several ways. Although random forces may also generate a noise spectrum which in turn perturbs the evolution of the trajectories, a more or less systematic variation of exogeneous systems parameters can usually be observed.

In social sciences, the empirical data base is often in a rather bad state. Limited time series of data, and considerable uncertainties of the values of the data must therefore always be taken into account.

The combination of uncertain data, namely external noise, via an 'active' environment with 'true' effects via nonlinear interactions of subsystems, may lead to irregular structures, difficult to understand and to interprete. Under such considerations the question of possibilities and limitations of the predictability of any socio-economic development is therefore of crucial interest.

The understanding of not only the 'coarse grained' structure of an economic system but of the interactions of its different subsystems requires a cautious treatment of each step of model building.

It is the aim of this work:
- to show how a stochastic theory of interrelated decisions of groups of agents can be used to establish a dynamic disequilibrium theory on the macro-level starting from general considerations on the micro-level,
- to show how constraints can be introduced within this theory,
- to demonstrate and to test under what conditions and/or circumstances well known results from (equilibrium) economics can be obtained,
- to show that the consideration of interrelated decisions of agents - in other words

the introduction of self-reinforcing mechanisms (synergy effects) - may result in a different evolution of the economy,
- to build a theoretical framework which can easily be extended and generalized without losing its general structure,
- to simulate and discuss by means of examples the dynamics of several economic model systems.

The paper is organized as follows: In Section 4.2 the stochastic basis of this economic disequilibrium theory is introduced. In Section 4.3 the stochastic framework of this theory is considered. The decisions of groups of agents in an economic system are treated via a dynamic decision model. Self-reinforcing mechanisms, in other words correlated decisions, can be handled via this approach. We derive a dynamic system of equations for the evolution of demand, supply and prices. As a limiting case this procedure provides well established results known from equilibrium considerations. In Section 4.4 we consider as an example the case of two commodities. In order to be specific, we simulate in Section 4.5 the evolution of demand supply and prices for the two decision models introduced. Possible extensions of the work are briefly discussed in an outlook, Section 4.6.

4.2 THE STOCHASTIC BASIS OF THE ECONOMIC DISEQUILIBRIUM THEORY

A model of economic growth or dynamic adjustment has to be explicit about the market dynamics of firm behaviour. Firm dynamics originates in innovative or imitative behaviour and in the price and quantity adjustments of agents when their mutually inconsistent plans are confronted in markets (Eliasson, 1989).

Specific institutions and mechanisms make it possible for the individual agent or a sector to continue functioning when their plans cannot be fulfilled. Following the argumentation of R.Day (1984)

Retail stores function as inventories mediating the flow of supplied and demanded commodities without the intervention of centralized coordination or of complicated and time-consuming market bargaining and negotiational procedures. Banks and other financial intermediaries regulate the flow of purchasing power among uncoordinated savers and investors, and mediators the flow of credits and depts that facilitate intertemporal exchanges without simultaneous bartering of goods. Ordering mechanisms with accompanying backlogs and variable delivery delays together with inventory fluctuations provide a flow of information that facilitates adjustment to disequilibria in commodity supplies and demands. Insurance and other transfer schemes such as unemployment compensation place resources in the hand of agents who without them would have no viable course of action.

In a first step we neglect spatial effects and develop a formal micro-based theory of behaviour in disequilibrium economics. The outlined theory is generic, so that different dynamic advantage functions (decision functions) and appropriate constraints can be introduced and tested without any change of the underlying structure of the mathematical formulation.

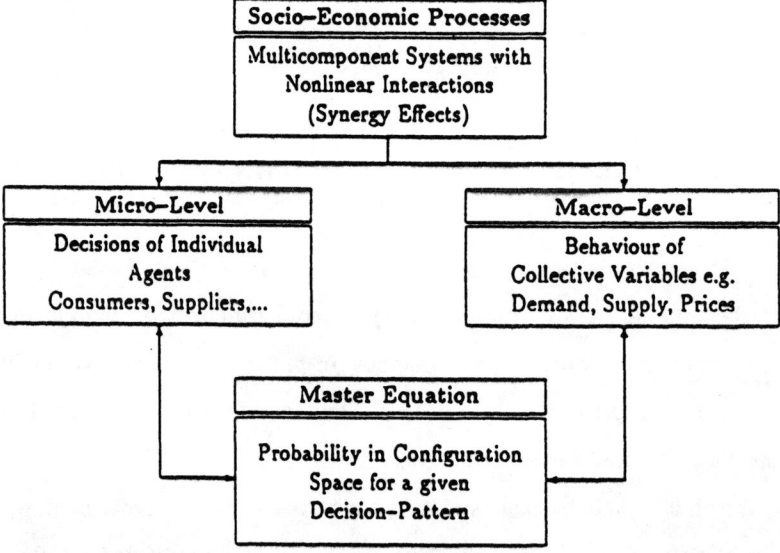

Figure 1: The master equation point of view: the relationship between the micro- and macro-level in decision processes.

The master equation formalism (Weidlich and Haag, 1983; Haag, 1989) provides an adequate framework for the dynamic modelling of a system which consists of many interacting micro-units (subsystems). Thus both levels of consideration namely the micro-level and the macro-level are linked via the master equation (see Figure 1)

4.2.1 The decision behaviour of different groups of agents: the micro-state of the economy

We consider different groups of agents P_α ($\alpha = 1, 2, ..., A$) in the economy. The individual agent is denoted by $I_\alpha^{(i)}$, ($i = 1, 2, ..., I$), where I is the number of individual $I_\alpha^{(i)}$ agents of subpopulation α. The different decision strategies of the individual agents with respect to their choice set $x_\alpha^{(i)} = \{x_{\alpha 1}, x_{\alpha 2}, ..., x_{\alpha J}\}^{(i)}$ taking into account certain constraints $N_\alpha^{(i)}$ constitute the starting point of our analysis. J denotes the number of mutually exclusive alternatives. It is assumed that there are a finite number and discrete bundles of commodities. Each individual agent tends to choose the option which is considered most desirable given his preferences and the attributes of each alternative as seen by the agent. In other words, it is assumed that an agent (e.g. consumer, producer) measures the desirability of each alternative $x_\alpha^{(i)}$ by a function $Z_{\alpha(x)}^{(i)}$.

Table 1. Different Agents of the Economic System

agent	decisions about	decision strategy	controlled variable
$I_\alpha^{(i)}$	$x_\alpha^{(i)}$	$\max_x Z_\alpha^{(i)}$ for $N_\alpha^{(i)} = 0$	distribution $\vec{x}$
consumers	commodities	utility optimization	demand $\vec{D}$
firms	prod. activities	profit optimization	supply $\vec{S}$
firms or retailers	prices of goods and services	profit optimization	price level $\vec{P}$

The corresponding Lagrangian function $L_\alpha^{(i)}$ for the derivation of the optimal solution reads

$$L_\alpha^{(i)} = Z_\alpha^{(i)} + \lambda_\alpha^{(i)} \cdot N_\alpha^{(i)}(\vec{x}) \qquad (4.1)$$

where the first-order conditions

$$\frac{\partial L_\alpha^{(i)}(\vec{x})}{\partial x_{\alpha j}} = 0 \tag{4.2}$$

are necessary for an optimum. In Table 1 we consider an economy consisting of 3 kinds of individual agents: consumers, whose decisions control the demand D of different goods, firms, which are responsible for the supply of goods and services S, as well as retailers (or firms) whose decisions fix the prices P.

Decisions of the agents have to be made under uncertain conditions, at least with respect to several characteristics of the commodities or services, and the structure of the market.

We consider a large number of consumers I (price-takers) who have a certain income budget. J is the number of desirable commodities on the market. Consumers must pay for the goods they want and because of their limited budgets, consumers who want to get greatest satisfaction have to make choices. We consider the response of each consumer and firm to an arbitrary (disequilibrium) price-configuration. Because the economy is in general out of equilibrium firms make non-zero profits (or losses) and these must be allocated among the consumers in determining their incomes.

A plausible decision strategy of the different agents consists of following the maximal gradient in phase space, in other words, consumers are searching for the maximal utility and firms are searching for the maximal profit, given certain constraints. The corresponding Lagrangian $L(\vec{x})$ can be seen as a dynamic advantage function (Haag, 1989). The rate of change between different alternatives is assumed to depend on differences of those dynamic advantage functions:

$$\max \{ L(\vec{x} + \vec{k}) - L(\vec{x}) \}, \tag{4.3}$$

or in case of slowly varying $\frac{\partial L}{\partial x_i}$ compared to $(\vec{x} + \vec{k}) - (\vec{x})$, in other words, if it is justified to replace the difference by the derivative this yields:

$$\max \{ \vec{k} \cdot \text{grad } L(\vec{x}) \}. \tag{4.4}$$

A reasonable form for the corresponding rate of change of the conditional probability per unit time, the 'transition rate' from a state $\vec{x} \rightarrow (\vec{x} + \vec{k})$ reads

$$w_t(\vec{x} + \vec{k}, \vec{x}) = v \cdot e^{L(\vec{x}+\vec{k}) - L(\vec{x})} \tag{4.5}$$

or alternatively using (4.4)

$$w_t(\vec{x} + \vec{k}, \vec{x}) = v \cdot e^{\vec{k} \operatorname{grad} L(\vec{x})}, \tag{4.6}$$

where the time-scaling parameter $v > 0$ denotes the flexibility of an individual to adopt different choices.

4.2.2 The macro-state of the economy

The macro-state of the economy is described by the socio-configuration (Haag, 1989).

$$\begin{aligned}
\vec{x} &= (\vec{D}, \vec{S}, \vec{P}) \\
&= (\vec{D}^{(1)}, ..., \vec{D}^{(I)}; \vec{S}^{(1)}, ..., \vec{S}^{(R)}; \vec{P}^{(1)}, ..., \vec{P}^{(R)}) \\
&= (D_{j=1}^{i=1,r=1}, D_1^{1,2}, ..., D_1^{1,R}, D_2^{1,1}, ..., D_2^{1,R}, ..., D_J^{1,R}, D_1^{2,1}, ..., \\
&\quad S_{j=1}^{r=1}, S_1^2, ..., S_1^R, ..., S_J^R, P_{j=1}^{r=1}, P_1^2, ..., P_1^R, P_2^1, ..., P_J^R)
\end{aligned} \tag{4.7}$$

here $\vec{D}^{(i)}$ is the 'demand vector' of an individual consumer (i), $\vec{D}$ is the total demand vector, $\vec{S}^{(r)}$ is the 'supply vector' of an individual firm, $\vec{S}$ is the total supply vector, $\vec{P}^{(r)}$ is the 'commodity price vector' of producer r, and $\vec{P}$ is the total price vector.

Of course demand, supply and prices have to fulfil certain constraints and properties which are not discussed in the following, but are always guaranteed.[1]

4.3 THE STOCHASTIC FRAMEWORK

Decision processes are stochastic, since we are interested in the probability $p(\vec{x}, t)$ that a certain (socio-)configuration is realized at time t. Of course, $p(\vec{x}, t)$ must satisfy at all times the probability normalization condition:

$$\sum_{\vec{x}} p(\vec{x}, t) = 1. \tag{4.8}$$

The master equation is the equation of motion describing the full dynamics in probabilistic terms. Thus the link between the micro-level and the macro-level is provided by this stochastic equation. For further details and specific applications of the master equation see Weidlich and Haag (1983), and Haag (1989). It reads:

$$\frac{dp(\vec{x}, t)}{dt} = \sum_{\vec{k}} \{w_t(\vec{x}, \vec{x} + \vec{k}) p(\vec{x} + \vec{k}, t) - w_t(\vec{x} + \vec{k}, \vec{x}) p(\vec{x}, t)\} \tag{4.9}$$

where the sum on the right hand side of (4.9) extends over all $\vec{k}$ with nonvanishing configurational transition rates $w_t(\vec{x}, \vec{x} + \vec{k})$, and $w_t(\vec{x} + \vec{k}, \vec{x})$, respectively. In configuration space the maximum of $p(\vec{x}, t)$ represents the most probable supply-, demand-, and price-configuration.

The total transition rate $w_t(\vec{x} + \vec{k}, \vec{x})$ is composed of contributions representing different economic processes:

$$\begin{aligned}
w(\vec{x} + \vec{k}, \vec{x}) &= \sum_{i=1}^{I} \sum_{j,k=1}^{J} w_{j,k}^{D,i} (\vec{D}^{(i)} + \vec{k}_D, \vec{S}, \vec{P}; \vec{D}^{(i)}, \vec{S}, \vec{P}) \\
&+ \sum_{r=1}^{R} \sum_{j,k=1}^{J} w_{j,k}^{S,r} (\vec{D}, \vec{S}^{(r)} + \vec{k}_S, \vec{P}; \vec{D}, \vec{S}^{(r)}, \vec{P}) \\
&+ \sum_{i=1}^{I} \sum_{j=1}^{J} w_{j+}^{D,i} (\vec{D}^{(i)} + \vec{k}_{D+}, \vec{S}, \vec{P}; \vec{D}^{(i)}, \vec{S}, \vec{P}) \\
&+ \sum_{i=1}^{I} \sum_{j=1}^{J} w_{j-}^{D,i} (\vec{D}^{(i)} + \vec{k}_{D-}, \vec{S}, \vec{P}; \vec{D}^{(i)}, \vec{S}, \vec{P}) \\
&+ \sum_{r=1}^{R} \sum_{j=1}^{J} w_{j+}^{S,r} (\vec{D}, \vec{S}^{(r)} + \vec{k}_{S+}, \vec{P}; \vec{D}, \vec{S}^{(r)}, \vec{P}) \\
&+ \sum_{r=1}^{R} \sum_{j=1}^{J} w_{j-}^{S,r} (\vec{D}, \vec{S}^{(r)} + \vec{k}_{S-}, \vec{P}; \vec{D}, \vec{S}^{(r)}, \vec{P}) \\
&+ \sum_{r=1}^{R} \sum_{j=1}^{J} w_{j+}^{P,r} (\vec{D}, \vec{S}, \vec{P}^{(r)} + \vec{k}_{P+}; \vec{D}, \vec{S}, \vec{P}^{(r)}) \\
&+ \sum_{r=1}^{R} \sum_{j=1}^{J} w_{j-}^{P,r} (\vec{D}, \vec{S}, \vec{P}^{(r)} + \vec{k}_{P-}; \vec{D}, \vec{S}, \vec{P}^{(r)})
\end{aligned} \tag{4.10}$$

The terms $w_{j,k}^{D,i}$ refer to changes in the demand-configuration due to decisions of consumer (i), from good k to good j. On the supply side, the change in the output distribution of firm (r) for production of good j instead of good k is described by the term $w_{j,k}^{S,r}$. The birth and death rates $w_{j+}^{D,i}$, $w_{j-}^{D,i}$, and $w_{j+}^{S,r}$, $w_{j-}^{S,r}$ describe a possible extension or reduction in the stock of demand and supply. For example, the introduction of a new product on the market can be represented by a birth process. Birth and death processes $w_{j+}^{P,r}$, $w_{j-}^{P,r}$, are further used in order to describe the increase or decrease in price level of good j, respectively.[2]

In (4.10) we neglected simultaneous changes in $\vec{D}, \vec{S}, \vec{P}$ to the total transition rate. This assumption is well justified, since in general different agents are responsible for different economic sectors.

There is a *cyclic coupling* between causes and effects, since the decisions of each group of agents depends on the decision pattern of the other groups.

4.3.1 The macro-dynamics of the economy

By substitution of the transition rates (4.10) into the master equation (4.9), the evolution of the probability distribution $p(\vec{x},t)$, starting from a given initial distribution $p(\vec{x}, 0)$ can be calculated. Of course, this procedure requires detailed information about the explicit structure of the transition rates, or in other words, about all economic choice strategies.

A fundamental advantage of this method consists in its generic structure. Different assumptions regarding transition rates in e.g. price dynamics, consumer behaviour or choice strategies of entrepreneurs can be tested without losing the general underlying structure of the mathematical formulation of the economy.

The tremendous amount of information contained in the configurational probability $p(\vec{x}, t)$ requires a less exhaustive description, e.g. in terms of macro equations describing the most probable (or the mean) behaviour of the economic system.

In order to derive equations of motion on the macro-level (in other words for the mean decision behaviour of the agents of the economy), we use the definition of the mean value of the socio-configuration

$$\vec{\bar{x}}(t) = \sum_{\vec{x}} \vec{x}\, p(\vec{x}, t) \tag{4.11}$$

and multiply (4.9) with $\vec{x}$ and sum up over all possible configurations. However, it is reasonable to assume that the mean value $\vec{\bar{x}}(t)$, of the socio-configuration practically coincides with the realized demand-, supply-, and price-configuration at time t.

If it is justified to assume that the probability distribution is a well behaved, sharply peaked distribution, the approximate relation

$$\overline{f(\vec{x}, t)} \approx f(\vec{\bar{x}}, t) \tag{4.12}$$

holds and a self-contained system of non-linear differential equations can be derived.

$$\begin{aligned}
\frac{d\bar{D}_j}{dt} &= \sum_{i=1}^{I} (\sum_{k=1}^{J} (w_{j,k}^{D,i}(\vec{D}, \vec{S}, \vec{P}) - w_{k,j}^{D,i}(\vec{D}, \vec{S}, \vec{P})) + w_{j+}^{D,i} - w_{j-}^{D,i}) \\
\frac{d\bar{S}_j}{dt} &= \sum_{r=1}^{R} (\sum_{k=1}^{J} (w_{j,k}^{S,r}(\vec{D}, \vec{S}, \vec{P}) - w_{k,j}^{S,r}(\vec{D}, \vec{S}, \vec{P})) + w_{j+}^{S,r} - w_{j-}^{S,r}) \\
\frac{d\bar{P}_j}{dt} &= \sum_{r=1}^{R} (w_{j+}^{P,r}(\vec{D}, \vec{S}, \vec{P}) - w_{j-}^{P,r}(\vec{D}, \vec{S}, \vec{P}))
\end{aligned} \tag{4.13}$$

However, the tremendous amount of information still contained in (4.13) requires an even less exhaustive description. Assuming homogeneous populations of agents, in other words we introduce 'representative' agents (consumers, entrepreneurs) characterized by e.g.

$$\sum_{i=1}^{I} w_{j,i}^{D,i} \approx I\, w_{j,k}^{D} \tag{4.14}$$

the dimension of the system of equations can tremendously be decreased. In the following we shall always use this assumption.

The mean value equations (4.13) belonging to the master equation (4.9) may have one or multiple stationary states and it depends on the initial conditions which of those stationary points will be approached. This means that the economic system may, (depending on the values of the trendparameters, which on the other hand are influenced by the socio-economic environment) approach one unique or one out of several possible equilibrium states. In case of time dependent parameters it may also be, that the control parameter of the system passes a so called critical value and the macro-equations of the system become unstable. However, if the economy had previously adapted to the now unstable equilibrium, it will suddenly evolve into a new stable state or a new dynamic mode e.g. a limit cycle or a chaotic attractor (Haag, Weidlich, Mensch, 1987). Such radical variations of the macrostate of the economy, denoted as phase transitions, are accompanied by an enhancement of fluctuations which are caused on uncertainties in the decision process of the individual economic agents.

In order to simplify the notation the bar on the mean values will be omitted e.g., $\overline{D_j(t)} = D_j(t)$.

The evolutionary equations of the economy become fully explicit, by inserting the individual transition rates (4.10), with (4.6) in (4.13). In the following we consider a specific form of the Lagrangian on the demand side (consumer behaviour) and the supply side (entrepreneurs) in order to be able to simulate and discuss by means of examples the dynamics of a specific model economy. Of course, the economic system under consideration can easily be generalized or extended to other more realistic decision strategies of the economic agents.

4.3.2 The Langrangian function on the demand-side

The law of demand states a relationship between the quantity of a good that consumers intend to buy, other things being equal, and the price of that good. We assume the following Lagrangian function

$$L_D(\vec{D}, \vec{P}) = U(\vec{D}) + \lambda \cdot N(\vec{D}, \vec{P}) = \sum_{j=1}^{J} a_j D_j^{b_j} + \lambda \cdot (\sum_{j=1}^{J} D_j P_j - u) \qquad (4.15)$$

where λ is a Lagrange parameter taking into account the limited budget u of the 'representative' consumers, and $U(D)$ is a strictly convex utility function (Dolan, 1985, Lancaster, 1966), where $0 < b_j < 1$ is the elasticity of demand, and $a_j > 0$ a positive parameter.[3] However, since the Lagrangian acts as a decision function in the transition rates, e.g. (4.6), the parameter a_j describes synergy effects (self-reinforcing processes). Obviously the Lagrangian (4.15) can easily be improved, e.g. by introducing quality aspects (Haag, Wunderle, 1990).

In Figure 2, a typical Lagrangian function for the case of two commodities within a given budget is shown.

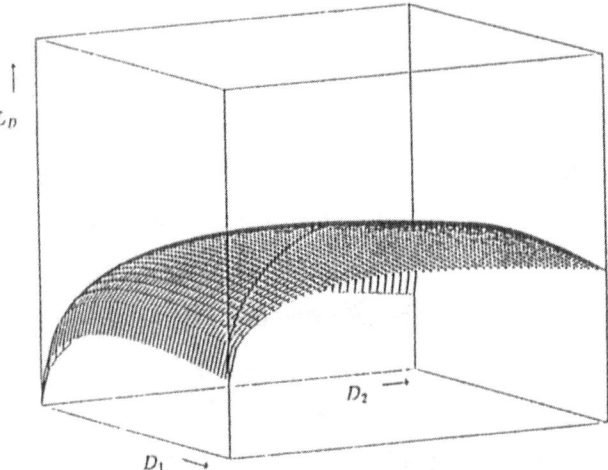

Figure 2: Lagrangian function for two commodities (limited budget). The limitation of the budget is indicated by the line on the surface. The parameters are listed in Section 4.5.

This Lagrangian (4.15) can be seen as the 'driving force' on the demand side.

4.3.3 The Langrangian function on the supply-side

On the supply-side we assume the following Lagrangian (profit function):

$$L_s(\vec{S}, \vec{P}) = \sum_{j=1}^{J} S_j P_j - \left[F + \sum_{j=1}^{J} (c_j S_j^3 + d_j S_j^2 + e_j S_j) \right] \quad (4.16)$$

where F are fixed inputs, and $C_j(S) = (c_j S_j^3 + d_j S_j^2 + e_j S_j)$ is the specific total-variable-cost curve (reverse-S shape) assumed[4] (Dolan, 1985). This means we assume a market structure of perfect competition. To obtain a reverse-S shaped total-variable-cost curve we choose: $c_j > 0$, $e_j > 0$, $d_j^2/3c_j$, $d_j < 0\$$. Figure 3 represents a typical profit function for the two commodity case.

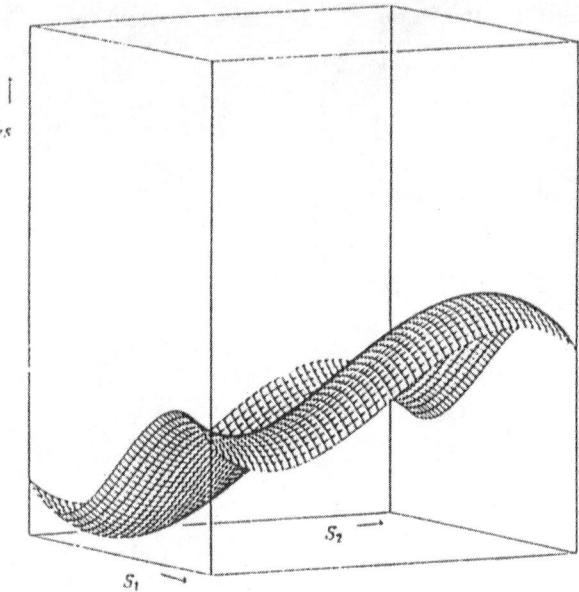

Figure 3: A typical profit function for two commodities. The parameters are listed in Section 4.5.

This Lagrangian (4.16) is the 'driving force' (decision function) on the supply side.

4.4   THE CASE OF TWO COMMODITIES

In this section we consider as an example an economic system assuming the following properties:

- A polypol is considered, consisting of a (homogeneous) population of suppliers and a (homogeneous) population of consumers.
- All agents are price-takers.
- The consumers have (all the same) limited budgets.
- The initial conditions for all consumers are identical, the same is true for the subpopulation of suppliers.
- The decision strategy of consumers is utility optimization; the suppliers are used to profit optimization.
- There are two completely substitutable goods on the market.

Inserting (4.16), (4.15) in the transition rates (4.6), (4.10) yield for the evolutionary equations of motion (4.13) on the macro-level

$$\frac{dD_1}{dt} = w^D_{1,2} - w^D_{2,1} + w^D_{1+} - w^D_{1-} \tag{4.17}$$

$$\frac{dD_2}{dt} = w^D_{2,1} - w^D_{1,2} + w^D_{2+} - w^D_{2-} \tag{4.18}$$

$$\frac{dS_1}{dt} = w^S_{1,2} - w^S_{2,1} + w^S_{1+} - w^S_{2-} \tag{4.19}$$

$$\frac{dS_2}{dt} = w^S_{2,1} - w^S_{1,2} + w^S_{2+} - w^S_{2-} \tag{4.20}$$

$$\frac{dP_1}{dt} = w^P_{1+} - w^P_{1-} \tag{4.21}$$

$$\frac{dP_2}{dt} = w^P_{2+} - w^P_{2-} \tag{4.22}$$

where transition rate

$$w^D_{1,2} = v_{1,2} \cdot e^{\left(\frac{\partial L_D}{\partial D_1} - \frac{\partial L_D}{\partial D_2}\right)} \cdot D_2^\gamma \tag{4.23}$$

refer to changes in the demand configuration due to decisions by consumers to choose good 1 instead of good 2, and

$$w^D_{2,1} = v_{2,1} \cdot e^{\left(\frac{\partial L_D}{\partial D_2} - \frac{\partial L_D}{\partial D_1}\right)} \cdot D_2^\gamma \tag{4.24}$$

vice versa. The parameters $v_{1,2}$, $v_{2,1}$ describe the flexibility of consumers to adapt to a disequilibrium situation. The increase or decrease of the stock of good j are described by the birth and death rates

$$w^D_{j+} = v_{j+} \cdot e^{\frac{\partial L_D}{\partial D_j}} \cdot D_j^\gamma \tag{4.25}$$

and

$$w^D_{j-} = v_{j-} \cdot e^{-\frac{\partial L_D}{\partial D_j}} \cdot D_j^\gamma \tag{4.26}$$

respectively, where according to (4.15) the derivative of the Lagrangian function

$$\frac{\partial L_D}{\partial D_j} = a_j b_j D_j^{b_j - 1} + \lambda P_j, \tag{4.27}$$

has to be inserted. The ν_{j+}, ν_{j-} are time-scaling parameters.

On the supply side, the transition rates for a shift in production $2 \rightarrow 1$ is assumed

$$w^S_{1,2} = \mu_{1,2} \cdot e^{\left(\frac{\partial L_S}{\partial S_1} - \frac{\partial L_S}{\partial S_2}\right)} \cdot S_2^\gamma \tag{4.28}$$

where the inverse process $1 \rightarrow 2$ reads

$$w^S_{2,1} = \mu_{2,1} \cdot e^{\left(\frac{\partial L_S}{\partial S_2} - \frac{\partial L_S}{\partial S_1}\right)} \cdot S_1^\gamma \tag{4.29}$$

The corresponding birth and death rates read

$$w^S_{j+} = \mu_{j+} \cdot e^{\frac{\partial L_S}{\partial S_j}} \cdot S_j^\gamma \tag{4.30}$$

and

$$w^S_{j-} = \mu_{j-} \cdot e^{-\frac{\partial L_S}{\partial S_j}} \cdot S_j^\gamma \tag{4.31}$$

respectively, with

$$\frac{\partial L_S}{\partial S_j} = P_j - [3c_j S_j^2 + 2d_j S_j + e_j]. \tag{4.32}$$

The $\mu_{i,j}$, $\nu_j\pm$ are time-scaling parameters as well.

The transition rates corresponding to an increase or decrease in the price level of good j read

$$w^P_{j+} = \kappa_j \cdot \frac{D_j}{D_j + S_j} \cdot P_j^\gamma \qquad (4.33)$$

and

$$w^P_{j-} = \kappa_j \cdot \frac{S_j}{D_j + S_j} \cdot P_j^\gamma \qquad (4.34)$$

This means that on the macro-level the *excess demand* ($D_j - S_j$) determines the rate of change of the prices:

$$\frac{dP_1}{dt} = \kappa_1 \left(\frac{D_1 - S_1}{D_1 + S_1}\right) P_1^\gamma \qquad (4.35)$$

$$\frac{dP_2}{dt} = \kappa_2 \left(\frac{D_2 - S_2}{D_2 + S_2}\right) P_2^\gamma \qquad (4.36)$$

The equations (4.17)-(4.22) become fully explicit by insertion of the transition rates (4.23) to (4.34). The solution of the corresponding highly nonlinear differential equations yields the dynamic development of the economy $D_j(t)$, $S_j(t)$, $P_j(t)$, given its initial state $D_j(0)$, $S_j(0)$, $P_j(0)$.

The parameter γ, $\gamma = 0$ or 1 corresponds to two different assumptions (two different models) concerning the decision strategy of agents.

The case $\gamma = 1$

The transition rates for $\gamma = 1$ are considered in Haag (1989). The justification of those rates is based upon the assumption that the flow of agents (e.g. number of consumers who decide to use commodity i instead of j per time unit) from one alternative to another is proportional to the number of agents who were already using the former alternative (e.g. the alternative D_j). Moreover, this parameter $\gamma = 1$, leads to a stationary state of the equations of motion on the macro-level, which can be characterized as a multinomial logit model.[5]

However, the stationary state of the macro-dynamics ($\gamma = 1$), namely the market equilibrium of demand, supply and prices is fixed via a transcendential system of equations. Because of the nonlinearities, multiple equilibria may exist and it depends on the history and exogeneous boundary conditions which of those equilibria points will be realized. It is worthwhile emphasizing that those points of market equilibria in general do not coincide with market clearing prices and a simultaneous optimal demand- and supply configuration. One reason for this discrepancy are self-reinforcing mechanisms (synergy effects), in other words correlated decisions of agents Arthur, (1988). As a consequence, the equilibrium point of a single agent (an individual) does not coincide with one of those stationary states on the macro-level.

Independent of the initial conditions and for all parameters (even completely unrealistic ones) positivity of demand, supply and prices is satisfied.

The case $\gamma = 0$

It can easily be seen that those micro-based decision strategies involved in the transition rates of the master equation for $\gamma = 0$ lead to a disequilibrium theory which provides all well established results known from equilibrium considerations. This statement holds if the system is at its stationary state and the economy is closed. The advantage of $\gamma = 0$ is that the stationary state on the micro-level corresponds to the stationary state on the macro-level. The reason for this is based on the neglect of correlated decisions (all agents are considered as independent of each other). A further disadvantage of the evolutionary equations for the case $\gamma = 0$ is that the positivity of demand, supply and prices is not always guaranteed. It depends on the initial conditions and the parameters chosen whether or not those requirements are fulfilled. However, in case of 'realistic' parameters positivity can always be assumed.

4.5 SOME SIMULATIONS

In this section we discuss a few numerical simulations for the case of two commodities. The parameters used in the simulations are listed in Table 2. The results shown in Figures 5 to 10 correspond to the set of dynamic equations (4.17) to (4.22). We assume the same total-variable-cost function for both types of commodities ($e_1 = e_2 = 53.6$; $d_1 = d_2 = -0.216$; $c_1 = c_2 = 0.0024$; $F = 10$). The corresponding profit function is shown in Figure 3. The fixed parameters of the utility function read ($a_2 = 1$; $b_1 = b_2 = 0.5$). The Lagrangian for the consumers ($a_1 = 1$; $u = 5000$) is represented in Figure 2. It is reasonable to assume that the speed of adjustment on the demand side is considerably higher than on the supply side.

Table 2. Parameters used for Simulations

v_{ij}	$v_{i,\pm}$	μ_{ij}	$\mu_{i,\pm}$	κ_i	γ	a_1	u	Figure
500	500	0	0.5	1	0	1.00	5000	4
500	500	0	0.5	20000	0	1.00	5000	5a,5b,5c
50	50	0	1	5	0	1.00	2800	6a,6b
50	50	0	1	5	1	1.00	2800	7
50	50	0	1	5	1	1.00	2500	8
50	50	0	1	5	1	1.05	2500	9
50	50	0	.1	5	1	1.20	2500	10
50	50	0	1	5	1	1.50	2500	11

In Figure 4 an over damped oscillation into the new globally stable equilibrium sets in. The stationary state corresponds to the profit and utility maximizing solution. Since the speed of adjustment of supply, demand and prices is different, the appearance of an excess demand is quite natural.

The parameters of the system are chosen completely symmetric with respect to goods 1, 2. Therefore, the stationary states for $\gamma = 1$ and $\gamma = 0$ are the same. However, the relaxation times of these two cases differ.

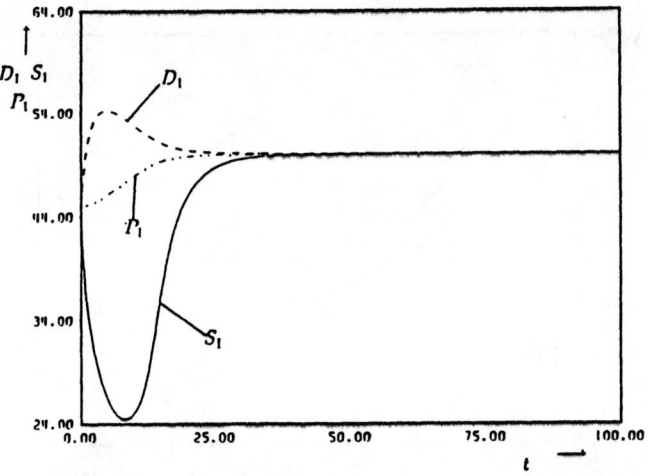

Figure 4. Supply S_1, demand D_1, and price P_1 as a function of time ($\gamma = 0$)

In Figures 5a to 5c, extreme flexibility parameters of the prices ($\kappa_1 = \kappa_2 = 20000$) are assumed. As recently mentioned (Schumann, 1987), the disequilibrium between supply S_j and demand D_j for a given price P_j can be considered as a usual phenomenon (Otani, El-Hodiri, 1987, Laux, 1982), even in a completely free market. Incomplete information is one proposed explanation of such non-market clearing prices. However, since the time scales of changing prices κ_i are certainly quite different from those of changing production processes μ_{ij}, $\mu_{i,\pm}$, and the time scales of changing consumers behaviour ν_{ij}, $\nu_{i,\pm}$ differ also considerably from the latter two, we consider the occurrence of an excess demand ($D_j - S_j$) as a dynamic phenomenon. Moreover, we expect that the prices are almost adiabatically adjusted to the current disequilibrium situation (Haag, 1990).[6]

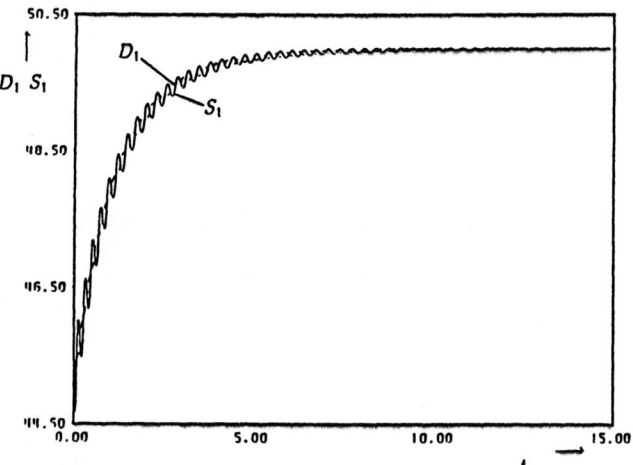

Figure 5a. The dynamics of supply S_1 and demand D_1 in the long-run

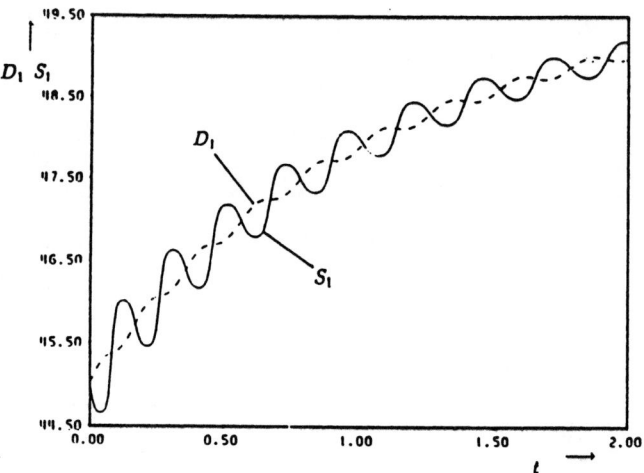

Figure 5b: The dynamics of supply S_1 and demand D_1 in the short-run

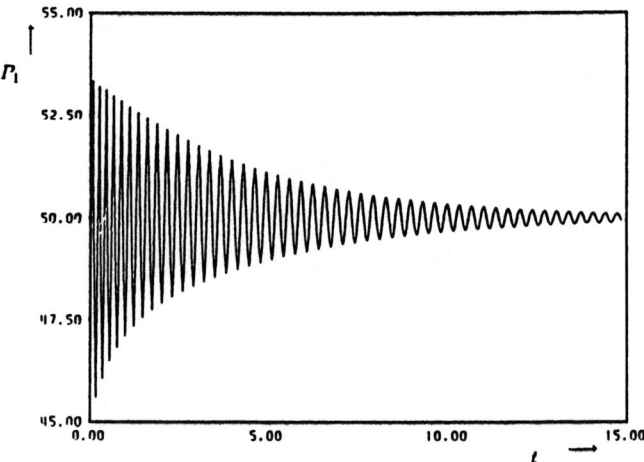

Figure 5c: The dynamics of the price of good 1, P_1, in the long-run

In Figures 5a and 5b the disturbed economy relaxes towards its new globally stable equilibrium, characterized by utility and profit optimizing groups of agents. However, 'fast' oscillations on demand and supply are superimposed the global evolution. The disequilibrium price dynamics, Figure 5c, exhibits a slowly damped oscillation.

In Figures 6a and 6b for $\gamma = 0$, and a medium price flexibility (see Table 1) a limit cycle appears, or in other words, we observe a stable oscillatory behaviour in prices, demand and supply.

For the same set of parameters and the same initial conditions, except $\gamma = 1$, this limit cycle is transfered into a stable focus and a slowly damped oscillation sets in as shown in Figure 7.

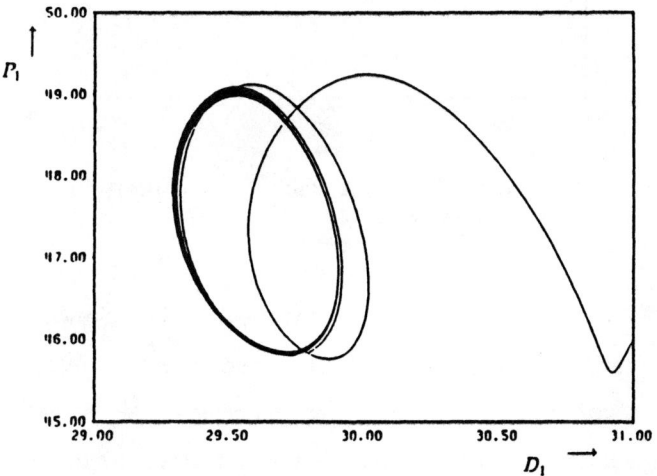

Figure 6a: The phase diagram for D_1, P_1. A stable limit cycle is approached for $\gamma = 0$

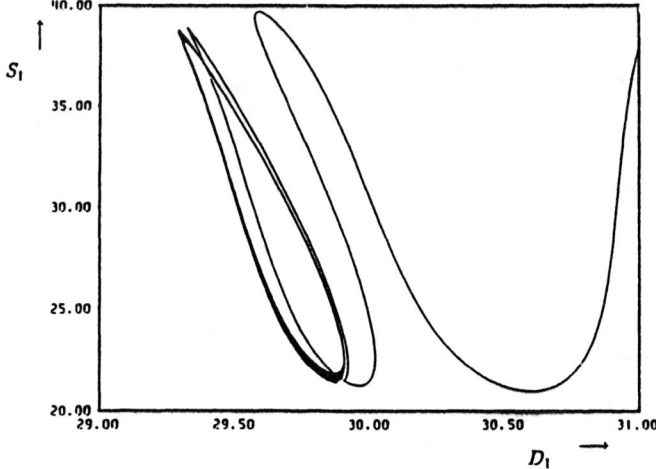

Figure 6b: The phase diagram for D_1, S_1. A stable limit cycle is approached (same parameters as in Figure 6a)

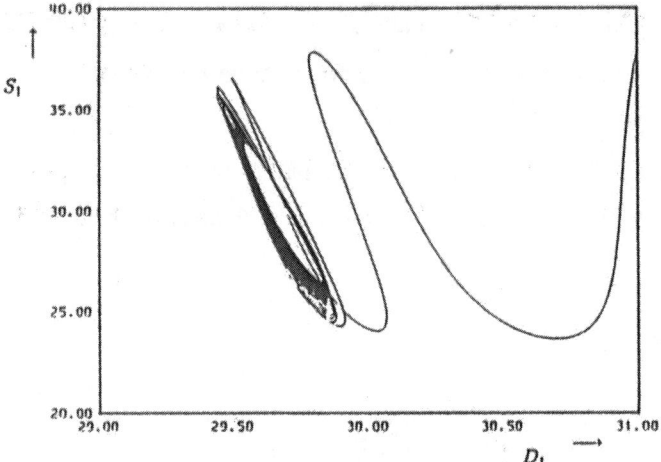

Figure 7: The phase diagram for D_1, P_1. A stable focus appears for $\gamma = 1$, all other parameters are the same as in Figures 6a, 6b

In Figures 8 to 11 a different scenario is represented. The chosen symmetry with respect to good 1 and good 2 is broken by a continuous increase of the attractivity of good 1, compared with good 2, all other parameters being equal. Increasing a_1, above a critical value a_{1c}, leads to a change in the pattern and/or the nature of critical points and may result in a complicated dynamics of the economy. In the following figures the temporal variation of the demand, $D_1(t)$, is depicted. Of course, all other variables, e.g. S_2, show a similar dynamic behaviour.

In Figure 8 a general trend with superimposed fast oscillations can be identified.

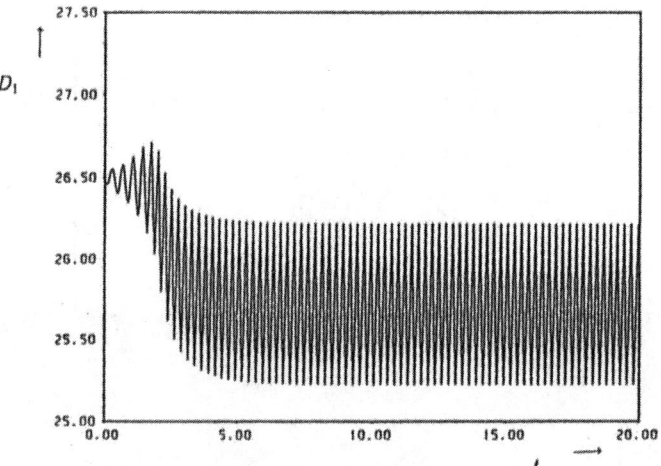

Figure 8: Demand $D_1(t)$, parameter $a_1 = 1.00$

Figure 9 can be interpreted as a 'phase transition' from one dynamic mode into another one. This interesting dynamic phenomenon results in the occurrence of a 'transient frequency doubling'. Therefore, the evolution of the economics shows some similarities to an intermittent transition to chaos.

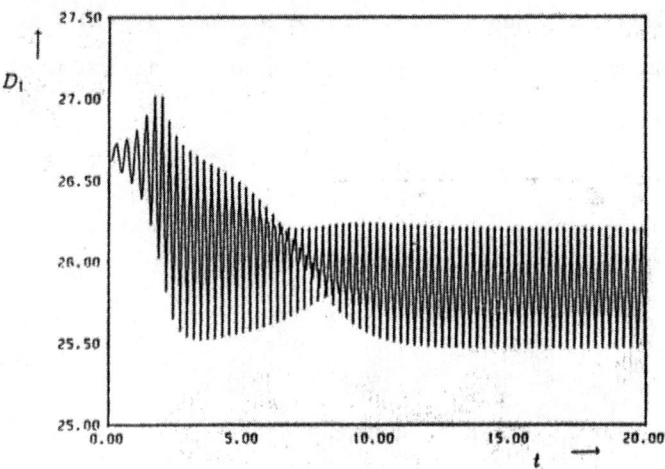

Figure 9: Demand $D_1(t)$, parameter $a_1 = 1.05$

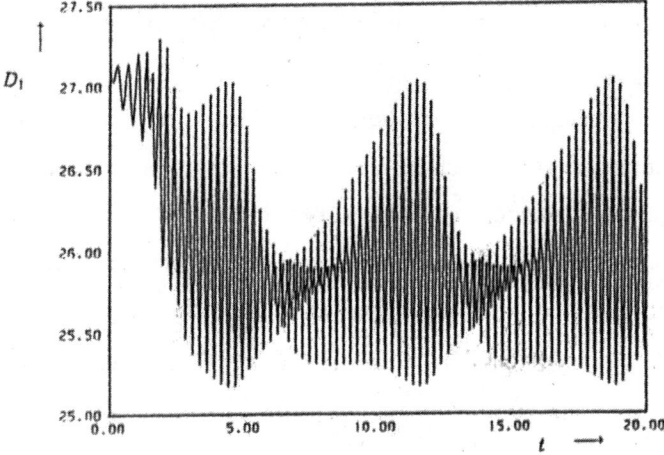

Figure 10: Demand $D_1(t)$, parameter $a_1 = 1.20$. Long-waves and short-waves in the economy can be identified

In Figure 10, a regular dynamic structure appears, characterized by quite different time scales. Phases of 'intermittent behaviour' are alternating with periods of enhanced and damped oscillations. Long-waves in the economy ($T_{long} >> T_{short}$ can clearly be distinguished from short-waves (fast oscillations in demand, supply and prices).

A further increase of the synergy parameter, a_1, Figure 11, results in a sudden breakdown of the long-term cycle. The frequency of the short-term oscillations remains almost unaffected. This is a common feature of many feedback systems. A very strong autocatalytic reaction mechanism (strong self-reinforcing forces) may lead to a reduction of the variiety of possible dynamic structures.

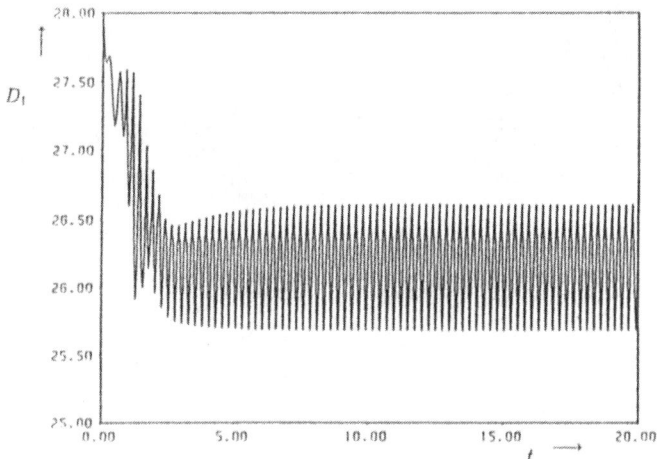

Figure 11: Demand $D_1(t)$, parameter $a_1 = 1.50$. The breakdown of the long-wave

Moreover, further simulations have shown that the economic system under consideration is sensitive to external perturbations, within a certain parameter space. In the case of time dependent parameters it may happen that a 'control parameter' passes a so called 'critical value' and the system becomes unstable. However, if the system had previously adapted to this now unstable equilibrium or unstable dynamic mode, it will suddenly rush into a new stable equilibrium state or a new dynamic mode e.g. a limit cycle or chaotic state.

4.6 OUTLOOK

The simulations presented show a variety of possible evolution patterns even for a simple economic system. The performed link between the micro- and macro-level of consideration builds a bridge for a better justification of dynamic disequilibrium models, belonging to different economic fields of application. The theory is generic. Therefore, the theoretical framework can easily be extended and generalized without losing its general structure.

The introduction of spatial aspects in this theory is crucial in the case of a possible application to space-time phenomena, such as innovation diffusion. To indicate where and how technological change should be introduced into this stochastic theory and the discussion of the implications of doing so will also be treated in forthcoming work.

REFERENCES

Arthur, B.W., 1988, **Self-Reinforcing Mechanisms in Economy**, in Santa Fe Institute Studies in Complexity, Addison-Wesley, Mass.

Batten, D., J. Casti, B. Johansson, (eds.), 1985, **Economic Evolution and Structural Adjustment**, Springer-Verlag, Berkeley.

Barentsen W., P. Nijkamp, 1993, Non-Linear Dynamic Modelling of Spatial Interaction, **Environment and Planning B**.

Day, R., 1984, **Adaptive, Disequilibrium Dynamics of Urban-Regional Development**, paper prepared at the Netherlands Institute for Advanced Study, Wassenaar.

Dolan, E.G., 1989, **Econometrics**, The Dryden Press.

Eliasson, G., 1989, **Modeling Long-Term Macroeconomic Growth**, paper prepared for the OECD International Seminar on Science, Technology and Economic Growth, Paris.

Fischer, F.M., 1983, **Disequilibrium Foundations of Equilibrium Economics**, Cambridge University Press, Cambridge.

Haag, G., 1990, **Dynamic Decision Theory. Applications to Urban and Regional Topics**, Kluwer Verlag, Dordrecht.

Haag, G., 1990, A Master Equation Formulation of Aggregate and Disaggregate Economic Decision-Making, **Sistemi Urbani**, 1, pp. 65-81.

Haag, G., Weidlich, W., and Mensch, G., 1987, The Schumpeter Clock, (D. Batten, J.Casti and B.Johansson, eds.), **Economic Evolution and Structural Adjustment, Lecture Notes in Economics and Mathematical Systems**, Springer, Heidelberg, New York, 1987.

Haag, G., Wunderle, P., 1990, Quality Improvement and Self-Reinforcing Mechanisms in Economy, (B. Fuchssteiner, T. Lengauer, H. Skala, eds.), **Methods of Operations Research**, 60, pp. 709-722.

Lancaster, K., 1966, A New Approach to Consumer Theory, **J. Political Economy**, 74, pp. 132ff.

Laux, H., 1982, **Entscheidungstheorie I und II**, Springer-Verlag, Berlin, Heidelberg.

Otani, Y., El-Hodiri, M., 1987, **Microeconomic Theory**, Springer-Verlag, Berlin, Heidelberg.

Samuelson, P.A., 1948, **Foundations of Economic Analysis**, Harvard University Press, Cambridge.

Schumann, J., 1987, **Grundzüge der mikroökonomischen Theorie**, Springer-Verlag, Berlin, Heidelberg.

Weidlich, W., Haag, G., 1983, **Concepts and Models of a Quantitative Sociology. The Dynamics of Interacting Populations**, Springer Series in Synergetics, Vol. 14, Springer, Berlin, New York.

NOTES

1. For details see Haag, 1990.

2. We consider the price dynamics as a stochastic process. However, there are no individual agents assigned to this process.

3. Assuming (4.15) the first-order condition (4.2) satisfy the second law of Gossen. Therefore consumer equilibrium is a state of affairs in which a consumer cannot increase the total utility gained from a given budget by spending less on one good and more on another.

4. Of course, the total-variable-cost curve assumed in only one out of a set of short-run cost curves that can be constructed and used.

5. However, this requires further assumptions especially concerning the birth and death rates on the supply- and demand-side of the economy.

6. For market-clearing prices $D = S$, or in other words, if the prices are adiabatically eliminated (slaving principle) the 'fast' oscillations disappear.

CHAPTER 5
COMPLEX TRANSIENT MOTION IN CONTINUOUS-TIME ECONOMIC MODELS
Hans-Walter Lorenz

5.1 INTRODUCTION

The investigation of chaotic dynamical systems has been a major focus of research in economic dynamics during the last decade. In the fashion of several applied sciences, an emphasis has been put on the description of so-called *strange attractors*, i.e., sets of points to which trajectories starting in a neighbourhood of this set eventually converge but which are neither a fixed point nor a closed curve. Strange attractors have attracted the attention of many economists because the motion on such a set is characterized by a *sensitive dependence on initial conditions*: two trajectories starting at arbitrarily close initial points eventually diverge implying that the two generated time series display different frequencies and amplitudes. This sensitive dependence has been considered an indication of the potential impossibility of predicting economic time series.

Most interesting nonlinear economic models which appear as good candidates for chaotic systems do not fulfil the requirements of the few mathematical theorems available for higher-dimensional dynamical systems. The best that often can be done with a higher-dimensional nonlinear system is to perform careful numerical experiments with the system under investigation, i.e., to determine the dynamic behaviour for different values of the parameters and initial conditions. Unfortunately, the mere calculation of trajectories can support a possible non-stationarity of the generated time series or the lack of other simple kinds of regular behaviour but the presence of chaotic dynamics cannot be established by eye-sighted inspection of the series. Non-parametric statistical tools have to be applied to a generated time series in order to determine whether the system behaves in a chaotic fashion or not. However, tools like Lyapunov exponents and correlation dimensions are defined for trajectories moving on attractors and usually require large data sets. Thus, it is not untypical that 10000 or 20000 points on an attractor have to be taken into account before reliable statistical results can be obtained.

Obviously, this large number of data points constitutes a problem for economics. For example, when a single data point represents quarterly figures the numerical experiments cover time horizons of several thousands of years. Depending on the assumed measuring scales of the state variables and adjustment coefficients this time horizon might be smaller but usually the horizon is very long. Economic "laws" of motion do not belong to the class of eternal laws known from classical physics but undergo structural changes. In many cases the presence of economic evolution probably contradicts the simulation of dynamical systems for more than a few years because attention has to be paid to the fact that parameters like savings rates, population growth rates, number and quality of goods, number and size of firms, etc. change over time.

When an economic dynamical system is valid only for a limited time span the relevance of the motion on an attractor can be questioned. It might happen that the system under consideration undergoes a structural change before the attractor is reached. Thus, the *transient* motion of a dynamical system before it reaches an attractor constitutes the relevant part of the trajectory in many economic examples. Interestingly, the transient motion can be extremely complex even when the attractor itself represents a regular object like a stable fixed point or a limit cycle. There even exist examples of arbitrarily long and complex transient behaviour in systems without a strange attractor.

In the rest of the paper this transient motion in continuous-time dynamical systems is studied in a simple economic example from business-cycle theory.[1] A few mathematical preliminaries are outlined in Section 2. Metzler's (1941) business-cycle model with inventories is studied as a didactical example of the Sil'nikov scenario described in Section 3. The role of complex transient behaviour in economic dynamics is discussed in the concluding Section 4.

5.2   FORMAL PRELIMINARIES

This paper deals only with complex transient motion in continuous-time dynamical systems. Thus, the following outline concentrates on differential equations though similar concepts exist for the discrete-time case.[2]

Consider a deterministic, n-dimensional system of nonlinear, ordinary differential equations defined on an open set $W \subset \mathbf{R}^n$

$$\dot{x} = f(x); x \in W \subset \mathbf{R}^n. \tag{5.1}$$

An *attractor* of a dynamical system (5.1) is a set of points to which a set of initial points eventually converges, i.e.,

Definition 1: A compact set $A \subset W$ is called an *attractor* if there exists a neighbourhood U of A such that for all $x(0) \in U$ one has $x(t) \in U \, \forall \, t \geq 0$ and $x(t) \to A$ for $t \to +\infty$, i.e., A is the w-limit set of all initial points in U.

A *repeller* is defined by replacing $+\infty$ by $-\infty$ in Definition 1, i.e., it is a set of points from which initial points in the neighbourhood of the set diverge.

The set of initial points which are attracted by A may constitute the entire phase space or only a subset of this space. In the latter case it is useful to introduce the notion of the *basin of attraction* of A:

Definition 2: Let U denote a neighbourhood of an attractor A. The *basin of attraction* $B(A)$ is the set of all initial points $x(0) \in U$ such that $x(t)$ approaches A as $t \to +\infty$, i.e., $B(A) = \cup_{t \leq 0} \phi_t(U)$ with $\phi_t(U)$ as the flow generated by (5.1).

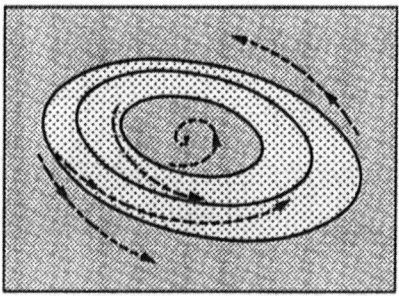

Figure 1   Regular Basin Boundaries

Different basins of attraction are bounded by their *basin boundaries*. When the attractors of a system like (5.1) are regular objects like fixed points or limit cycles, the basin boundaries typically are regular objects as well. An example of regular basin boundaries is outlined in Figure 1. The system is supposed to possess two attractors, namely the fixed point in the centre of the figure and an attracting limit cycle in the form of the middle ellipse. The dark-shaded area in the centre of the figure represents the basin of attraction of the fixed point; the light-shaded area represents the basin of attraction of the limit cycle. The inner ellipse constitutes the basin boundary between the fixed point and the limit cycle. The basin boundary of the limit cycle is formed by both the inner and outermost ellipses.

Basin boundaries do not necessarily have to form the regular and connected sets depicted in Figure 1. In the example to be introduced below the basins of attraction of two competing attractors consist of disconnected sets, the form of which can be very complicated. There exist dynamical systems with *fractal* basin boundaries, i.e., the dimension of the boundary is a non-integer number implying a fissured geometric shape.

Complex basin boundaries can be responsible for the emergence of complex transient motion. In the following, a transient motion will be called complex if it fulfils one of the following properties:

. The system possesses regular, non-chaotic attractors but the basin boundary is a complicated set. When the system starts in a neighbourhood of this set the possibly very long transient motion can be highly complicated. The transient motion may not be distinguishable from the motion on a strange attractor.

. The system possesses regular and/or chaotic attractors but the basins of attraction (with possibly fractal basin boundaries) are disconnected and spread over the phase space. While a trajectory converging toward one of the attractors can be rather regular the transient motion will be called complex when a precise knowledge of the initial values is required for the determination of the eventually approached attractor.

. Almost all initial points converge toward an attractor but a saddle-straddle trajectory exists which remains in a bounded region away from the attractor for a long time.

Usually, the second property is not denoted as a sort of complex transient motion. However, as economic theory should never make its statements dependent on absolutely precise numerical values of variables and parameters, the resulting fuzziness in the assumed initial values can imply a completely different convergence behaviour.[3]

The first property in the above-mentioned list represents the standard scenario for complex transient motion. The complicated set is usually a Cantor set, implying that a trajectory starting in this set can stay there forever and displays chaotic behaviour. When a trajectory starts in a neighbourhood of this set it can perform a chaotic behaviour before it eventually converges toward a regular attractor. The emergence of such a Cantor set can be illustrated with Smale's notion of the *horseshoe map* which will be outlined in the following.

Consider a continuous-time dynamical system of the form (5.1) and let n=3. Then it is possible to construct a two-dimensional Poincaré map on a cross section with the trajectory of the original 3D system. The Poincaré map displays all essential dynamic properties of the continuous-time 3D system. Consider the set of initial points in such a Poincaré map G located in the square region S in Figure 2.a. It is a basic property of chaotic dynamical systems that sets of initial points are stretched and contracted. Assume that the rectangle in 2.b is the result of the stretching and contracting in a first iteration step of the underlying map G. As the motion should be bounded a third component in the form of folding has to be considered; otherwise the stretching would rapidly lead to a convergence of one side of the rectangle toward infinity during the subsequent iterations. Assume that the folding takes place in such a way that the horseshoe-like shape in 2.b results. The figure has been drawn such that the bended horseshoe overlaps the original square, i.e., only a subset of the initial points contained in the square remain in this set after one iteration. Those elements of the horseshoe which overlap the square are leaving the set of points in the square. Figure 2.c shows the result of the second iteration of the original square. Stretching and contracting lead to the rectangle in 2.c. which is folded once again such that it overlaps the original square. The four shaded strips represent the initial points that succeeded in remaining in the square after two iterations. After k iterations, the appropriate figure would contain 2k vertical strips. By working backwards from 2.c to the original square it is possible to describe those points in the square S which have been mapped to itself after two iterations. The four horizontal strips in Figure 3.a represent those initial points which do not leave the square after two iterations.

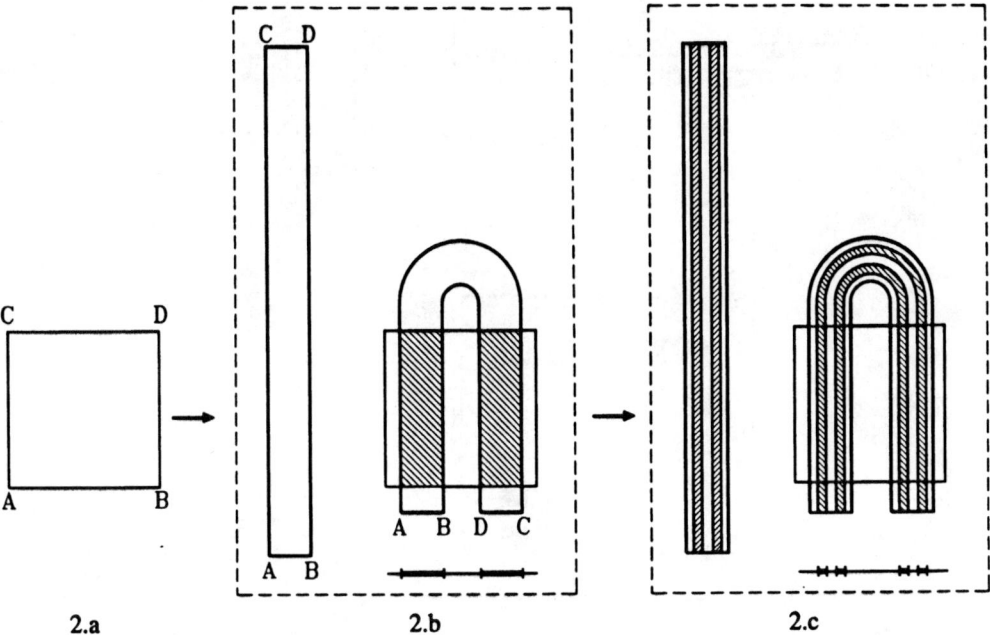

Figure 2   The Construction of a Horseshoe

The strips in Figure 3.a contain those points in S that will stay in S after two iterations, the strips in Figure 3.b contain those points in S that originated in S. The intersection of these strips thus describes the set of points that originated in S after two iterations and stay in S for two iterations. The intersection is depicted in Figure 4 for two iterations. When $k \to \infty$, the shaded rectangles shrink to an infinite number of points. The resulting set is invariant under the action of the dynamical system and is an example of a *Cantor set*.

It was demonstrated by Smale (1963) that the motion in this invariant set is chaotic (defined similarly as in Li/Yorke (1975)). However, the construction of the map uncovers that most initial points in the original square leave the square. Thus, orbits starting in a neighbourhood of the invariant set are dominated by it for a while before they eventually diverge from it.

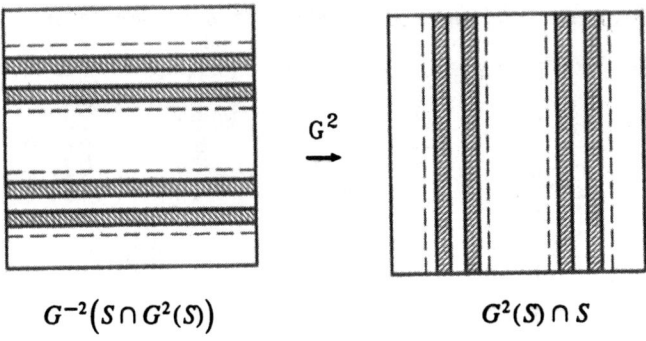

$G^{-2}(S \cap G^2(S))$ $G^2(S) \cap S$

Figure 3

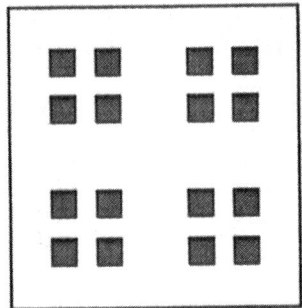

Figure 4 The Invariant Set in the Horseshoe Map

Such horseshoe maps can be detected in a variety of examples. A particular example was studied by Sil'nikov (1965).[4] If a three-dimensional system has a fixed point x* with one real eigenvalue and a pair of conjugate complex eigenvalues such that the real eigenvalue is absolutely larger than the real part of the complex eigenvalues, and if there is a homoclinic orbit for x*, then there exists a perturbed system with a homoclinic orbit and a countable set of horseshoes.

Figure 5 depicts a homoclinic orbit that is consistent with the Sil'nikov scenario. The orbit leaves the fixed point on the unstable manifold (determined by a positive real eigenvalue) and returns to the stable manifold (determined by the conjugate complex eigenvalues with negative real parts). The essential difficulty in applying Sil'nikov's result to a particular dynamical system consists in establishing the existence of a homoclinic orbit. There exist numerical algorithms for detecting their existence (cf. Beyn (1990)) but in the present context it is helpful to quote a result by

Arneodo/Coullet/Tresser (1981, 1982) and Coullet/Tresser/Arneodo (1979), namely that the dynamical system

$$\dddot{x} + a\ddot{x} + \dot{x} = f_\mu(x) \tag{5.2}$$

with a as a constant, fulfils the assumption of the Sil'nikov scenario for appropriate forms of the one-parameter family of functions $f_\mu(x)$. The forms of $f_\mu(x)$ which imply the Sil'nikov case include a logistic functional form and a piecewise-linear tent function.

The third-order differential equation (5.2) is particularly interesting for economic dynamics. Most economic dynamic models are framed in linear dynamical systems including constant terms. The constancy of these terms is usually a consequence of simplifying a given economic model. Once the constancy assumption is abandoned, it is often simple to find reasonable nonlinear replacements for the formerly constant terms. It turns out that rather weak assumptions regarding the involved nonlinear terms are sufficient to encounter the scenario investigated by Arneodo et al. In the following a very simple example of the emergence of an equation like (5.2) with a unimodal form of $f_\mu(x)$ will be discussed for illustrative purposes.[5]

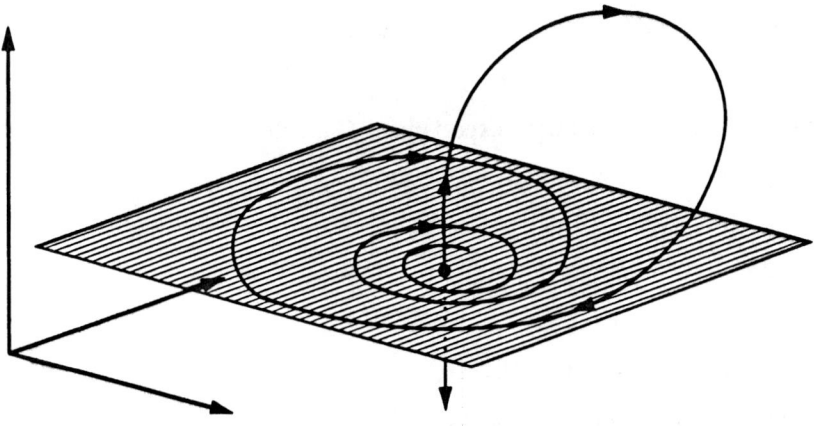

Figure 5   A Sil'nikov-type Homoclinic Orbit

5.3 TRANSIENT MOTION IN A CONTINUOUS-TIME BUSINESS-CYCLE MODEL

The following continuous-time model was first discussed by Gandolfo (1983) and represents a modification of an originally discrete-time version of Metzler (1941). As the model has been discussed elsewhere in different contexts, it will only be outlined in the following. The model describes a macro-economy in which production, Y, changes when the desired inventory stock deviates from the actual stock, i.e.,

$$\dot{Y} = \alpha(B^d(t) - B(t)), \qquad \alpha > 0, \qquad (5.3)$$

with $B^d(t)$ as the desired and $B(t)$ as the actual inventory stock at t. The actual inventory stock changes when there exists an excess demand or supply in the goods market, i.e.,

$$\dot{B}(t) = S(t) - I(t), \qquad (5.4)$$

with as $S(t)$ savings and $I(t)$ as investment. The desired inventory stock depends linearily on expected output, $Y^e(t)$, i.e.,

$$B^d(t) = kY^e(t), \qquad k > 0. \qquad (5.5)$$

While Metzler (1941) assumed static expectations, Gandolfo (1983) considered a modified extrapolative expectations hypothesis of the form

$$Y^e(t) = Y + a_1\dot{Y}(t) + a_2\ddot{Y}(t), \qquad (5.6)$$

with $a_1 > 0$ and $a_2 > 0$ as constants.

Combining the ingredients of the model and adjusting it to the fixed-point values Y*, B*, and B^{d*} yields[6]

$$\dddot{y} + A_1\ddot{y} + A_2\dot{y} = \beta(s(t) - i(t)), \qquad (5.7)$$

with A_1 and A_2 as constants containing the previous parameters, and y,s(t), and i(t) as the deviations of Y, S, and I from their fixed-point values Y^*, S^* and I^*, respectively.

In the following it will be assumed that $0 < A_1 < A_2$ and A_2 close to 1. Furthermore, $\beta(s(t) - i(t))$ is assumed to be a unimodal function of y, i.e., $\beta(s(t) - i(t)) = \beta y(d - y)$, $d > 0$. Equation (5.7) can be written as the following system of differential equations:

$$\dot{y} = x$$
$$\dot{x} = z \qquad\qquad (5.8)$$
$$\dot{z} = -A_2 x - A_1 z + \beta y(d - y)$$

with x and z as auxiliary variables.

System (5.8) possesses the two fixed points $(y^*, x^*, z^*) = (0,0,0)$ and $(\bar{y}^*, \bar{x}^*, \bar{z}^*) = (d,0,0)$. It can be shown (cf. Lorenz (1992b)) that the origin is a saddle with a positive real eigenvalue and a pair of complex eigenvalues with negative real part when the assumption $0 < A_1 < A_2$ holds. The second fixed point (d, 0,0) is locally asymptotically stable for $\beta < (A_1 A_2)/d$. For $\beta > (A_1 A_2)/d$, the real eigenvalue is negative and the real parts of the conjugate complex eigenvalues are positive. The last inequality is fulfilled in all numerical experiments described below.

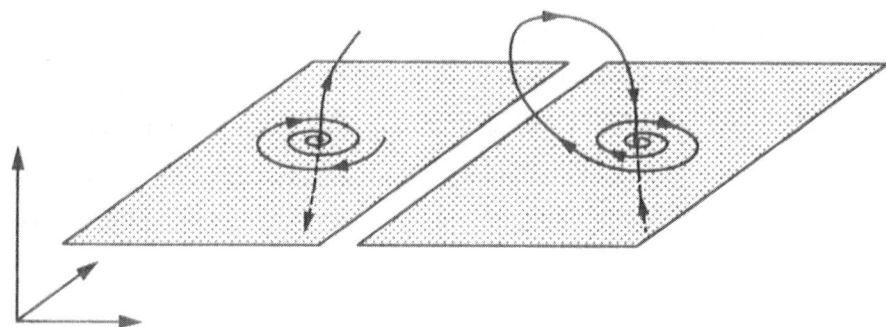

Figure 6   The Manifold Scenario in Equation (5.8)

The manifolds of the two fixed points are sketched in Figure 6. Both fixed points fulfil the local requirements of the Sil'nikov scenario, namely that a real eigenvalue λ, and a pair of conjugate complex eigenvalues γ, $\bar{\gamma} = \alpha \pm i\beta$ exist with the property $|\lambda| > |\alpha|$. The numerical experiments presented in the following sections indicate that the relevant oscillatory behaviour takes place in the neighbourhood of the fixed point (d,0,0) and that the homoclinic orbit connects the stable and unstable manifolds of this fixed point. The fixed point (0,0,0) is always located close to or on the basin boundary between a finite attractor and infinity. It should be noted that Figure 6 only vaguely describes the manifolds; the eigenvectors for particular parameter values may imply that the linear eigenspaces are slightly rotated into different directions.

The following three subsections deal with three typical forms of dynamic behaviour observable in (5.8). Provided the above mentioned inequalities are fulfilled similar results can be obtained for different parameter sets.

Strange "Attractors"

Figure 7 contains a projection of a trajectory generated by (5.8) onto the (y,x)-plane for a particular parameter set. A 3D presentation would not provide much more information because the trajectory generates a rather flat object with increasing z-values for decreasing values of y and x. It should be noted, however, that the trajectory spirals around the stable manifold of the fixed point (d,0,0) which itself has a slightly positive slope at this fixed point. In the 2D projection in Figure 7 this spiralling motion implies that the trajectory seems to spiral around different centres.

Figure 7 shows the trajectory of (8) for 55000 integrations. A larger number of integrations would imply that the trajectory filled the entire area already described by the trajectory in Figure 7. In fact, the trajectory seems to wander arbitrarily between inner and outer spirals. The Lyapunov spectrum is approximately $\lambda^L = (0.05, 0, -0.5)$, indicating that there exists a sensitive dependence of the motion on the object in Figure 7 on initial conditions. If chaos is defined in terms of the largest Lyapunov exponent, the trajectory in Figure 7 can thus be called a chaotic trajectory.

In the performed numerical experiments no indication for the convergence toward another attractor could be found. However, this does not exclude that the trajectory in Figure 7 represents an extremely long-lasting transient motion. Without the construction of the Poincaré map for the assumed parameter values and an inspection whether horseshoes overlap the original set of initial points no definitive answer on the nature of the convergence behaviour of the trajectory can be given. However, from a practical point of view of economics the observable long-lasting bounded motion qualifies the trajectory as attracted to the generated object in the figure for the admissible time span. The object can thus be called a strange "attractor" within the limited time in which the dynamical system can be considered a valid description of an actual phenomenon.

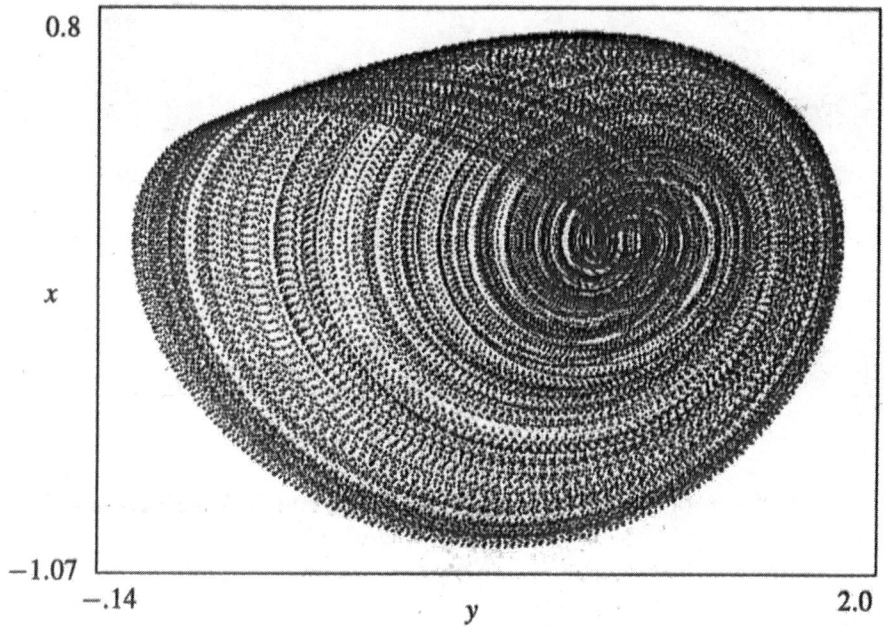

Figure 7 An "Attractor" of (8.8) in the (y, ẏ) plane; $A_1 = 0.4$; $A_2 = 0.95$; $\beta = 0.6$; $d = 1.3$; $T = 1\text{-}55000$.

For practical purposes it is useful to consider ± infinity an attractor. Figure 8 shows the basin of attraction of the object in Figure 7 and the basin of infinity. Initial points located in the white area converge toward the object in Figure 7 (with the above mentioned qualification); initial points in the shaded area converge toward minus infinity. The basin boundary appears to be regular. The numerical experiments do not indicate the presence of other attractors.

Changes in the parameters can destroy the shape of the object in Figure 7. For example, an increase in A_1 opens the central region around the fixed point (d,0,0) and eventually a closed double loop emerges. Further increases in A_1 lead to a regular limit cycle and finally to a stable fixed point. Before a single attracting closed loop emerges multiple attractors can be observed.

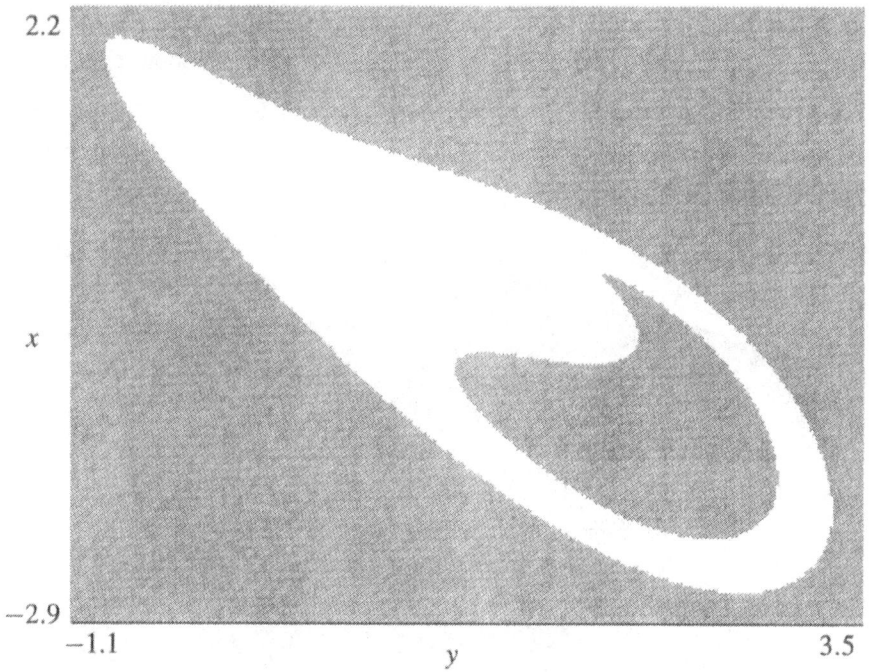

Figure 8 The Basin of Attraction of the Object in Figure 7 (White Region). The Grey-Shaded Areas Represent the Basin of Infinity.

Multiple Attractors

Figures 9 and 11 show the first 55000 phase points of the trajectories generated by (5.8) for a slightly increased value of A_1 and different initial conditions. The trajectory in Figure 9 spirals toward inner and outer regions of the generated object in a seemingly erratic fashion comparable to the trajectory in Figure 7. The Lyapunov spectrum is approximately $\lambda^L = (0.2, 0, -.4)$. While the largest Lyapunov exponent is smaller than in the previous case, no indication of a convergence toward zero during additional integrations exists. Thus, the trajectory in Figure 9 is (slightly) dependent on initial conditions and can be called chaotic.

As in Figure 7, no eventual divergence of the trajectory from the object in Figure 9 could be observed. Figure 10 shows the trajectory of (5.8) during integration steps

T=200001-255000. Although this time span is certainly beyond the admissible time span, the calculation allows to conclude that the object in Figures 9 and 10 is again a strange "attractor".

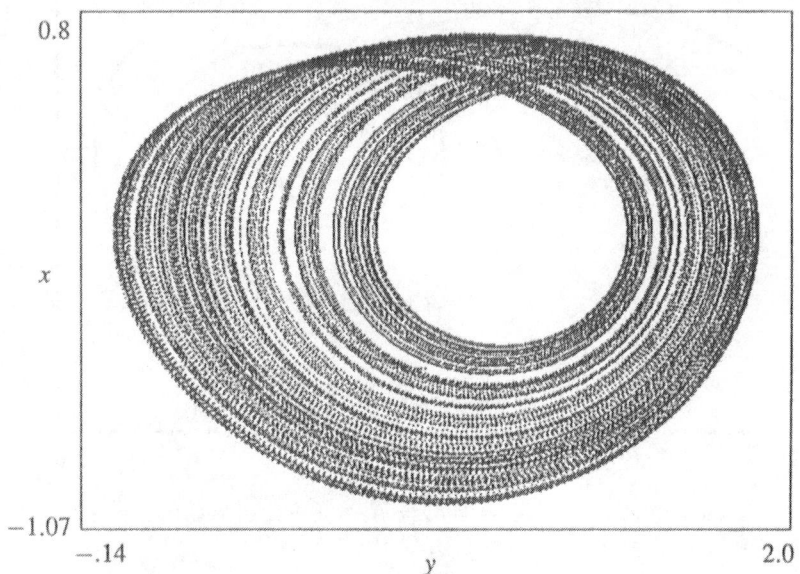

Figure 9   A trajectory of (8); $A_1 = 0.425$, $A_2 = 0.95$, $\beta = 0.6$, $d = 1.3$, $y(0) = 0.9$, $x(0) = -0.5$, $z(0) = 0.0$; T = 1-55000

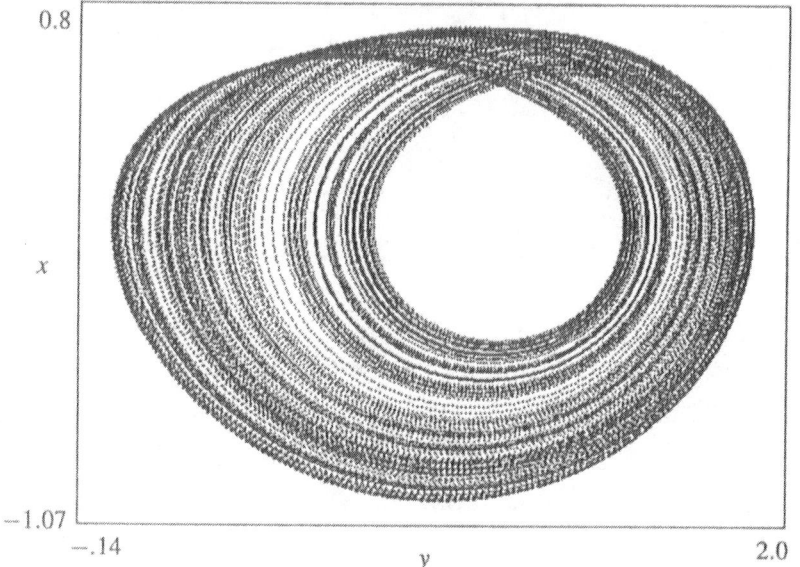

Figure 10   A Trajectory of (8); $A_1 = 0.425$, $A_2 = 0.95$, $\beta = 0.6$, $d = 1.3$, $y(0) = 0.9$, $x(0) = -0.5$, $z(0) = 0.0$; T = 200001-255000

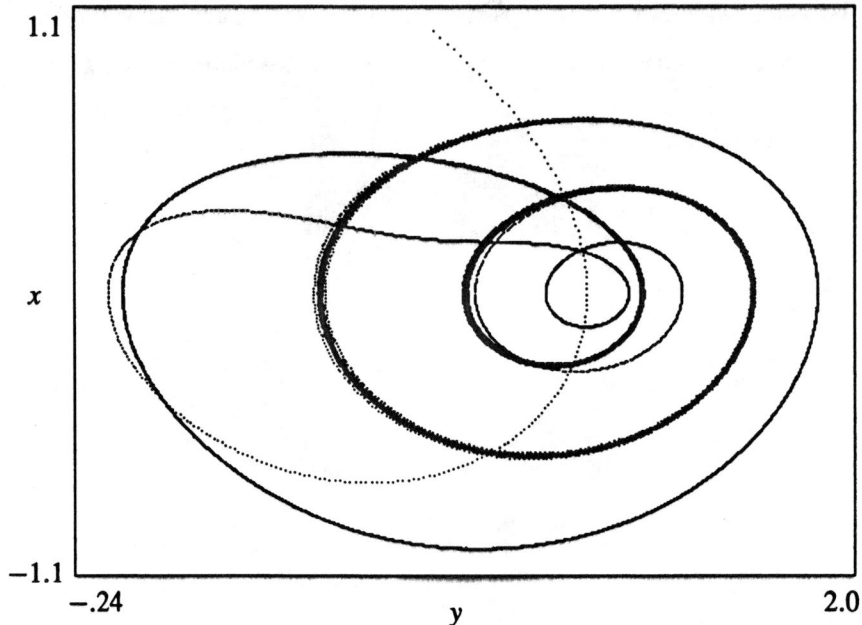

Figure 11 A Trajectory of (8); $A_1 = 0.425$, $A_2 = 0.95$, $\beta = 0.6$, $d = 1.3$, $y(0) = -0.8$, $x(0) = 1.9$, $z(0) = 0.0$; T = 51-55000

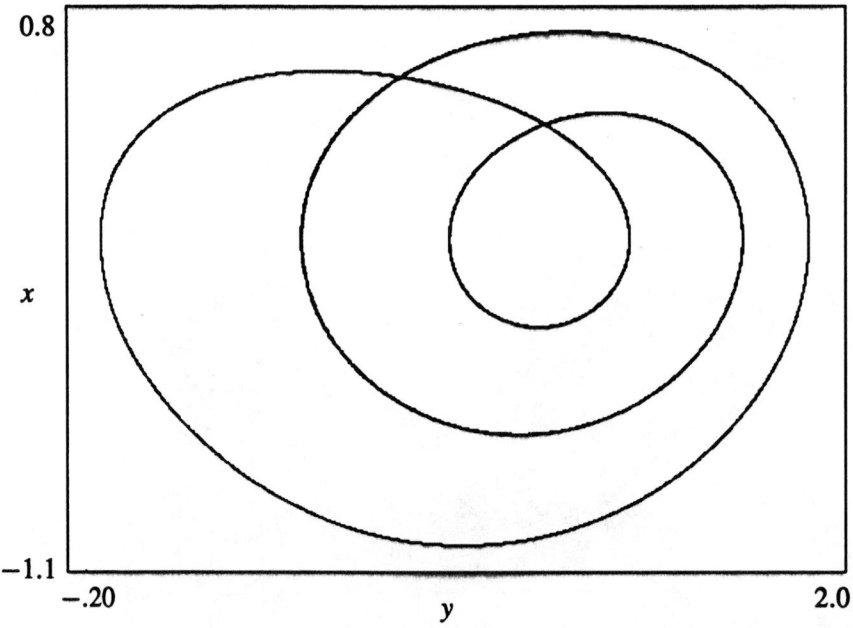

Figure 12 A Trajectory of (8); $A_1 = 0.425$, $A_2 = 0.95$, $\beta = 0.6$, $d = 1.3$, $y(0) = -0.8$, $x(0) = 1.9$, $z(0) = 0.0$; T = 10001-20000

For the same parameter set and particular sets of initial values a second attractor exists. Figure 11 shows the first 55000 points in phase space (the very first 50 points have been dropped so that the figures have a comparable scale). The trajectory converges toward the single closed curve forming a double loop around the fixed point (d, 0, 0). The exclusion of a considerable number of pre-iterates in Figure 12 uncovers that the closed curve can indeed be called an attractor (once again with the above mentioned qualifications).[7] The diameter of the ball around the fixed point described by this closed curve is slightly larger than the ball described by the trajectory in Figure 9.

Figure 13 shows the basins of the two attractors in the (y, x)-space. Initial points located in the white areas rapidly converge toward the attractor in Figure 9, initial points located in the black areas rapidly converge toward the attractor in Figure 12. The grey shaded areas represent initial points that belong to the basin of infinity. The basin boundary between the two finite attractors appears to have a fractal nature particularly in the neighbourhood of the basin of infinity.[8]

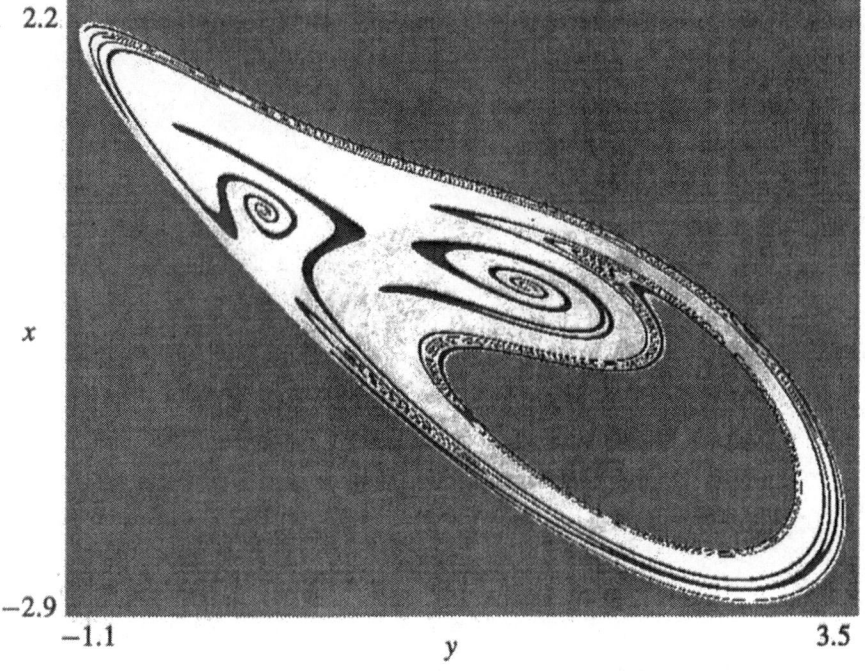

Figure 13   Basins of Attraction of the Attractor in Figure 7 (White Regions) and Figure 8 (Black Regions). The Grey-Shaded Areas Represent the Basin of Infinity.

The presence of these complicated basins of attraction qualifies the motion as a complicated transient motion in the sense of the previous definition. In certain regions of the phase space only a very precise knowledge of the initial conditions allows to predict whether the trajectory converges toward a regular, double-loop orbit or a slightly chaotic attractor.

A Saddle-Straddle Trajectory

In addition to parameter constellations implying the cases of single strange attractors (Figure 7) and multiple, bounded attractors (Figures 9 and 11) there exist examples of parameter values which are responsible for a divergence of the trajectories toward infinity for almost all initial values. Figure 14 shows several initial values, the trajectories of which remain in a bounded region comparable to the one described by the attractor in Figure 7 for some time. However, after a more or less longer lasting time span the trajectories enter the vicinity of the fixed point at the origin and converge rapidly toward infinity. No evidence for the existence of an attractor exists for this parameter constellation. In this sense the expression "almost all" has been used above.

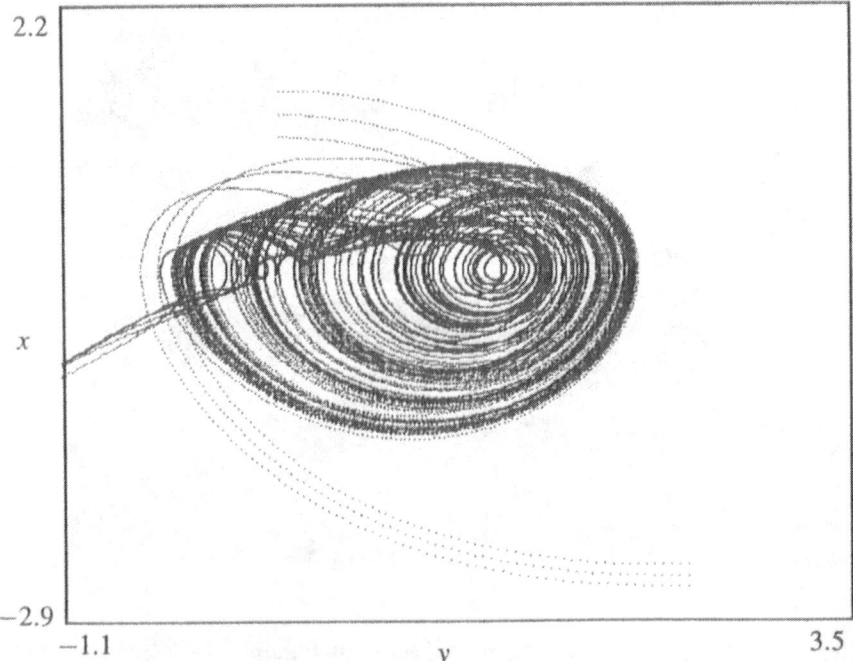

Figure 14   Transient Behaviour of Several Initial Points Before the Divergence Toward Infinity; $A_a = 0.425$, $A_2 = 0.95$, $\beta = 0.587$, $d = 1.5$.

Following the argument that a given dynamical system can be considered an accurate description of real life economic phenomena at most for a limited time span, the behaviour depicted in Figure 13 can be called a complex transient motion. The regular attractor to which trajectories eventually converge and which implicitly has been assumed in the discussion above only has to be replaced by the notion of an attractor at infinity. When the eventually diverging time series generated by this system are cut off a few integration steps before the trajectory leaves the bounded region and if the Lyapunov spectrum is calculated for the number of, say, first 2000 integration steps, it is possible to encounter Lyapunov spectra with entries typical for a chaotic attractor. Other initial values lead to a rapid divergence toward infinity without any positive entries in the spectrum. Of course, the number of integration steps is much too small to draw reliable conclusions on the dynamic behaviour of the underlying system when such a relatively small data set is available and the underlying dynamical system is not known. However, this scenario describes the standard situation encountered by an econometrician who usually has to deal with much shorter time series. Summarizing, even if almost all trajectories leave the region described by the various trajectories in Figure 14, some trajectories stay in that region for a considerable time span, the end of which might be longer than the actual expiration date of the economic dynamic model.

Although almost all trajectories converge toward infinity, there is nevertheless a particular trajectory that does not leave a bounded area and displays chaotic behaviour. The trajectory is defined on a *chaotic saddle*, i.e., an invariant set with chaotic properties which is not an attractor. A *saddle straddle trajectory* is an artificially generated trajectory produced in the following way.[9] As almost all initial points diverge toward infinity, an initial point that stays in a bounded, finite region can be found with probability zero. Instead, an initial point of the desired, bounded trajectory is found by considering two different initial values $y_a(0)$ and $y_b(0)$ which might be located in the basin of infinity. Both initial points and the points on a line connecting them will almost certainly leave the considered bounded region but the escape times of all points on the lines will differ. The two initial points are moved closer together such that the new connecting line contains a point with a larger escape time. This "straddling" is continued until both end points of the line have come together very closely. One of the two points is then considered the starting point of the desired trajectory. A second point on the desired trajectory is found by iterating both end points of the (now very short) line. As trajectories diverge, the resulting points $y_a(1)$ and $y_b(1)$ will be farther apart then the end points of the previous line. The two resulting points are then moved closer together in the same way as above. When they have approached sufficiently close, a second point on the desired trajectory has been determined, etc.

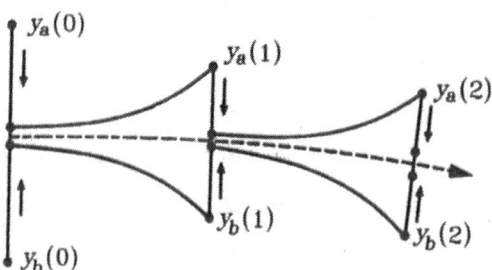

Figure 15   The Construction of a Saddle-Straddle Trajectory

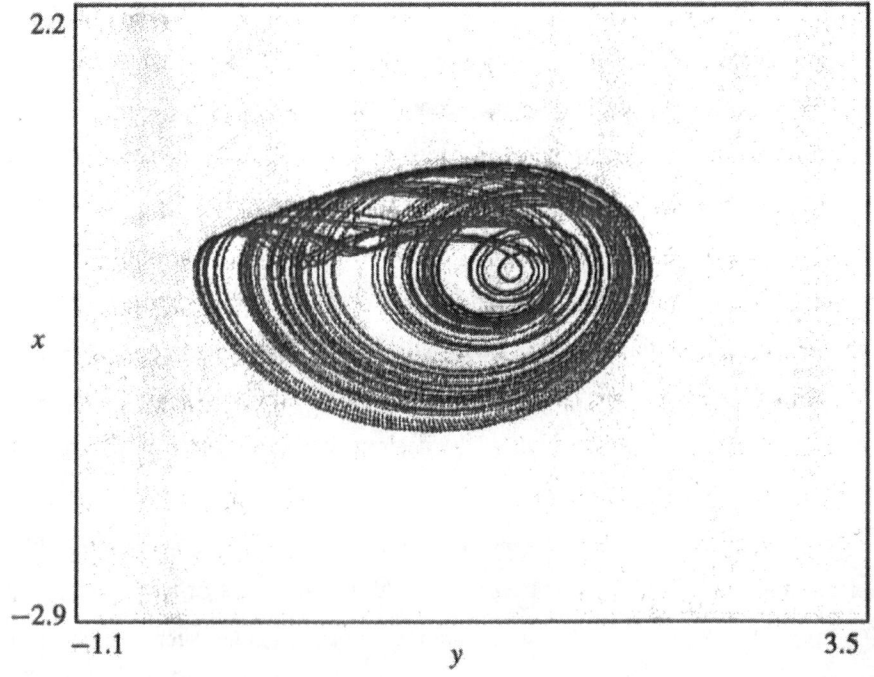

Figure 16   A Saddle-Straddle Trajectory of (8); $A_1 = 0.425$, $A_2 = 0.95$, $\beta = 0.587$, $d = 1.5$

Figure 16 contains such a saddle-straddle trajectory for the same parameter set as in Figure 14.[10] Initial points located precisely on this trajectory will stay on the trajectory forever. However, if a trajectory is calculated by iterating the underlying dynamical system with the standard integration techniques instead of the straddle method described above, the resulting trajectory will eventually diverge from the saddle-straddle trajectory because numerical errors can never be avoided. As the chaotic saddle is not an attractor, a trajectory slightly off the saddle trajectory will finally converge toward infinity, i.e., the attractor in the example of equation system (5.8). The time an arbitrary trajectory spends in the neighbourhood of the chaotic saddle may nevertheless be very long such that one may speak of a long-lasting, chaotic, transient motion.

5.4 THE RELEVANCE OF COMPLICATED TRANSIENT MOTION IN DYNAMIC ECONOMIC MODELS

The discussion of this simple business-cycle model with inventories has shown that complex transient motion can be observed in continuous-time models of economic dynamics. The three types of complex transient motion originate in different scenarios: 1) a parameter constellation implying a chaotic attractor (with the qualification that the attractor may constitute only an extremely long-lasting transient), 2) a parameter constellation with multiple attractors (including a chaotic attractor and a single closed curve) and a complicated, possibly fractal basin boundary, and 3) a constellation with a single attractor (infinity) and a chaotic saddle which is not an attractor. The nature of continuous-time dynamical systems requires the consideration of long time horizons, and purely chaotic or complex transient motion turns out to be essentially different from regular behaviour only after integrating the dynamical system many times. Besides, the transition from complex transient behaviour to a regular orbit is usually not as spectacular as in discrete-time systems where a sharp distinction between complex transient and regular oscillatory behaviour in the generated time series can be observed in many examples.[11] However, if the measuring scales of the variables and the dimension of (possibly only implicitly present) adjustment speeds are taken into account the typical number of iterations of economic dynamic models in the range of several thousands of iterations may cover only a few years. Thus, the assumed time spans in the calculations performed above do not *per se* constitute a purely academic exercise.

Nearly all models in nonlinear economic dynamics concentrate on attractors (excluding infinity) in the form of stable fixed points, limit cycles, or strange attractors. This seems to be a reasonable procedure when the concentration on attractors is justified with the observation that economies usually do not collapse in historic time, i.e., they remain in a bounded, finite region in phase space. However, this argument overlooks two possible types of dynamic behaviour in nonlinear dynamical systems:

1. There might exist one or several finite attractors of a dynamical system, some of which might be regular orbits (including fixed points). The researcher might be able to detect these attractors and might arrive at the conclusion that a motion on one of these finite attractors describes the actual motion of the economy which is modeled with the help of the dynamical system. However, such an attractor may never be reached in a time span allowed to be considered the validation time of the assumed dynamical system. When a complex and longer lasting transient period cannot be excluded, the simultaneous existence of a regular attractor and complicated motion in a bounded region do not necessarily contradict each other. More philosophically, the stability dogma of neoclassical economics with its concentration on stable fixed points does not inherently exclude complex phenomena like transient chaotic behaviour.

2. Economics traditionally distinguishes between stable and unstable states of an economic dynamic system. While it is certainly true that an economist's object is not the modelling of collapsing economies (at least in most cases), this distinction is actually irrelevant when the investigated system is nonlinear and when complex, longer lasting transient motion cannot be excluded. The example studied at the end of the previous section demonstrates that an economy described by a dynamical system like (8) can be completely unstable in the sense that almost all initial points converge toward infinity. An analysis concentrating on these final state would

describe the underlying model as economically unsatisfactory. Nevertheless, the trajectories might stay in a finite region for a considerable time span; the presence of structural change in actual economic systems might imply that a trajectory generated by a specific dynamical system comes to a halt before its trajectories diverge to infinity. The final point of this system serves as the initial value of a newly emerging dynamical system.

This discussion assumes that the validation time span of a dynamical system with fixed parameter values is limited. The question whether this assumption is justified depends on the specific economic phenomenon which is to be described by the model at hand. For example, business-cycle models that attempt to describe long cycles like Juglar or Kondratieff cycles and which assume, say, constant marginal rates of consumption or constant rates of money supply, should not be considered satisfactory models of economic evolution. At least, the parameters of a model should be allowed to vary from time to time. From a dynamical systems point of view the validity of a given dynamical system is thus restricted to the time span in which certain parameters can safely be considered constant. Other economic examples can be quoted in which parameter changes do not play an essential role. Short-term models in the economics of financial markets and auctions probably do not have to care much about structural changes in the underlying dynamical system (when it exists). Economic models that explicitly attempt to describe long-run phenomena are in principle good candidates for a limited validation time span.

Urban and regional economics deals with time spans which are usually beyond the time horizon of standard macroeconomic dynamic models. The phenomena which are to be described include the evolution of transportation systems, the migration of populations in urban areas, etc., i.e., phenomena that take place on time scales of several decades. When such a model assumes a constancy in parameter values which cannot safely be considered constant for a long time span, the mere investigation of the system's attractors should be treated with some care in the light of possibly long-lasting transient behaviour in the discussion of the simple example presented above. When parameters can be treated as constant over a long time span, the possible presence of a complicated transient behaviour before the eventual convergence toward a regular attractor does not exclude a seemingly erratic development of the system variables.

The Metzlerian business-cycle model has been used in the above presentation because some results known in the mathematical literature can directly be applied to this system. It has already been mentioned that a dynamical system like (5.2) can emerge in a variety of economic scenarios. However, the findings can probably be translated to other economic dynamical systems. The study of continuous-time dynamical systems is not untypical in urban economics, and a prominent example has been provided by Dendrinos (1986) who investigated a spatial employment model which can be framed in the well-known Lorenz equations. It has been known for some time that this three-dimensional system possesses homoclinic orbits (cf. Glendinning/Sparrow (1984) for details). The local properties of the fixed points are different from in the Sil'nikov scenario but it should be expected that similar types of complex transient motion emerge for particular parameter sets and initial values.

REFERENCES

Alligood, K.T. and J.A. Yorke, 1989, Accessible Saddles on Fractal Basin Boundaries, Mimeo University of Maryland.

Arneodo, A., P. Coullet, and C. Tresser, 1981, Possible New Strange Attractors with Spiral Structure, **Communications in Mathematical Physics 79**, pp. 573-579.

Arneodo, A., P. Coullet, and C. Tresser, 1982, Oscillators with Chaotic Behavior: An Illustration of a Theorem by Sil'nikov, **Journal of Statistical Physics 127**, pp. 171-182.

Beyn, W.-J., 1990, The Numerical Computation of Connecting Orbits in Dynamical Systems, **IMA Journal of Numerical Analysis 9**, pp. 379-405.

Coullet, P., C. Tresser, and A. Arneodo, 1979, Transition to Stochasticity for a Class of Forced Oscillators, **Physics Letters 72A**, pp. 268-270.

Delli Gatti, D., M. Gallegati and L. Gardini, 1991, Investment Confidence, Corporate Debt, and Income Fluctuations, Mimeo Urbino.

Dendrinos, D.S., 1986, On the Incongruous Spatial Employment Dynamics, (P. Nijkamp, ed.), **Technological Change, Employment, and Spatial Dynamics**, Springer-Verlag, Berlin-Heidelberg-New York, pp. 321-339.

Gandolfo, G., 1983, **Economic Dynamics: Methods and Models**, 2nd Edition, North-Holland, Amsterdam.

Glendinning, P. and C. Sparrow, 1984, Local and Global Behavior near Homoclinic Orbits, **Journal of Statistical Physics** 35, pp. 645-696.

Grebogi, C., E. Ott, and J.A. Yorke, 1987a, Crises, Sudden Changes in Chaotic Attractors, and Transient Chaos, **Physica 7D**, pp. 181-200.

Grebogi, C., E. Ott and J.A. Yorke, 1987b, Basin Boundary Metamorphoses: Changes in Accessible Boundary Orbits, **Physica 24D**, pp. 243-262.

Grebogi, C., E. Ott, and J.A. Yorke, 1987c, Chaos, Strange Attractors, and Fractal Basin Boundaries in Nonlinear Dynamics, **Science** 238, pp. 632-638.

Guckenheimer, J. and P. Holmes, 1983, **Nonlinear Oscillations, Dynamical Systems, and Bifurcations of Vector Fields**, Springer-Verlag, New York-Berlin-Heidelberg.

Kantz, H. and P. Grassberger, 1985, Repellers, Semi-Attractors, and Long-Lived Chaotic Transients, **Physica 17D**, pp. 75-86.

Li, T.Y. and J.A. Yorke, 1975, Period Three Implies Chaos, **American Mathematical Monthly** 82, pp. 985-992.

Lorenz, H.-W., 1992a, Multiple Attractors, Complex Basin Boundaries, and Transient Motion in Deterministic Economic Systems, (G. Feichtinger, ed.), **Dynamic Economic Models and Optimal Control**, North-Holland, Amsterdam, pp. 411-430.

Lorenz, H.-W., 1992b, Complex Dynamics in Low-Dimensional Continuous-Time Business Cycle Models, **System Dynamics Review** 8, pp. 233-250.

McDonald, S.W., C. Grebogi, E. Ott and J.A. Yorke, J.A., 1985a, Fractal Basin Boundaries, **Physica 17D**, pp. 125-153.

McDonald, S.W., C. Grebogi, E. Ott and J.A. Yorke, J.A., 1985b, Structure and Crisis of Fractal Basin Boundaries, **Physics Letters** 107A, pp. 51-54.

Mira, C., 1987, **Chaotic Dynamics - From the One-Dimensional Endomorphism to the Two-Dimensional Diffeomorphism**, Singapore World Scientific.

Medio, A., 1991, Continuous-time Models of Chaos in Economics, **Journal of Economic Behavior and Organization** 16, pp. 115-151.

Metzler, L.A., 1941, The Nature and Stability of Inventory Cycles, **Review of Economic Studies** 23, pp. 113-129.

Nusse, H.E. and J.A. Yorke, 1989, A Procedure for Finding Numerical Trajectories on Chaotic Saddles, Physica 36D, pp.137-156.

Sil'nikov, L.P., 1965, A Case of the Existence of a Countable Number of Periodic Motions, **Sov. Math. Dokl.** 6, pp. 163-166.

Smale, S., 1963, Diffeomorphisms with Many Periodic Points, in: Cairns, S.S., (ed.), **Differential and Combinatorical Topology**, Princeton Princeton University Press, pp. 63-80.

Wiggins, S., 1988, **Global Bifurcations and Chaos, Analytical Methods**, Springer-Verlag, New York-Berlin-Heidelberg.

Wiggins, S., 1990, **Introduction to Applied Nonlinear Dynamical Systems and Chaos**, Springer-Verlag, New York-Berlin-Heidelberg.

Yorke, J.A., 1991, Dynamics, **An Interactive Program for IBM Clones**, University of Maryland.

NOTES

1. A discussion of several economic examples of complex transient motion in one-dimensional, discrete-time systems is contained in Lorenz (1992a). A corporate-debt model by Delli Gatti/Gallegatti/Gardini (1991) implies complicated transient behaviour in a two-dimensional discrete-time map.

2. An extensive discussion of transient motion in discrete-time, one-dimensional endomorphisms and two-dimensional diffeomorphisms can be found in Mira (1987). A lot of detailed information can be obtained from Grebogi/Ott/Yorke (1987a,b,c), Kantz/Grassberger (1985), and McDonald/Grebogi/Ott/Yorke (1985a,b).

3. Such a dependence on initial conditions can, of course, be found in all dynamical systems with multiple attractors. The complexity mentioned above is due to the fact that in the described scenario the basins are formed by a large number of disconnected pieces with possibly complicated boundaries.

4. Compare Guckenheimer/Holmes (1983), pp. 319, Wiggins (1988), pp. 227ff., and Wiggins (1990), pp. 540ff., for details.

5. Compare Medio (1991) for a detailed analytic discussion of a similar, refined model. The following model is described in more detail in Lorenz (1992b) which also contains another example of the emergence of the Sil'nikov scenario in economic dynamics.

6. Cf. Lorenz (1992b) for details.

7. The calculation of the basin boundary straddle trajectory cf. Nusse/Yorke (1989)) uncovers that it possesses the same shape as the closed curve in Figure 12. However, this straddle trajectory is slightly shifted away from the closed curve in Figure 12.

8. The basins of the bounded attractors form two spirals in Figure 13. The right spiral emerges around the fixed point (d,0,0). The presence of the left spiral, however, is surprising because the fixed point (0,0,0) is not located in its centre. In the neighbourhood of the centre of this spiral no extraordinary behaviour could be detected.

9. Details are described in Nusse/Yorke (1989) and Yorke (1991).

10. The figure was generated with Yorke's (1991) Dynamics program.

11. Cf. Lorenz (1992a) for a few economic examples in one-dimensional and two-dimensional systems.

CHAPTER 6
ECONOMIC STRUCTURE AND NONLINEAR DYNAMIC DEVELOPMENT
Capital, Population and Knowledge
Wei-Bin Zhang

6.1 INTRODUCTION

Economic growth and development theory, like the economic system itself, is prone to severe cyclical fluctuations in popularity and interest. Important variables, such as population, capital and knowledge, appear and disappear cyclically in dynamic theory, according to social and economic circumstances, not to mention individual interest. For instance, Malthus' population theory, Ricardo's two-sector economic system and neo-classical growth theory have been emphasized by different economists in different cultures at different times. This also suggests possible existence of a general framework in which current macroeconomic models can be treated as special cases. It is obvious that such a synthesis, if possible, must be formulated in the form of a high dimensional nonlinear dynamic system. Due to the accumulated efforts of economists and, in particular, due to recent developments of nonlinear dynamic theory (e.g., Haken, 1977, 1983, Nicolis and Prigogine, 1977, Wilson, 1981, Weidlich and Haag, 1983, Rosser, 1991, Zhang, 1991, Dendrinos and Sonis, 1990), it appears that the timing is right for such a synthesis. The purpose of this study is to make an initial attempt in this direction - we try to build a macroeconomic development model, synthesizing Malthus' population theory, Ricardo's two-sector economic system and neo-classical growth theory.

The paper is organized as follows. Section 2 defines the basic model. Section 3 guarantees the existence of equilibria and provides stability conditions. Sections 4 and 5 analyze, respectively, the effects of changes in the government's research policy and the creativity of different activities upon the system. Section 6 studies how the system structure is affected by the parameters of the population dynamics. Section 7 shows the complicated impact of the savings rate upon the economic structure. Section 8 concludes the study. The appendix interprets stability conditions when the population is fixed.

6.2   THE BASIC MODEL

We consider an economic system consisting of two production sectors - agriculture and industry - and one university. The agricultural sector produces commodities for consumption. The industrial sector produces commodities either for consumption or for investment. The university makes contribution to knowledge growth. The university is a public sector in the sense that it is financially supported by tax income from the producers.

The total labor force, N, is distributed among the two production sectors and the university. Assume that the amount of available land, L, is given. Prices are measured in terms of the agricultural commodity. Let p(t) denote the price of the industrial commodity. We assume that the labor force, land and capital are homogeneous and that all the markets are perfectly competitive. This implies that the wage rate, w(t), the land rent, R(t), and the interest rate, r(t), are identical throughout in the whole economy.

agriculture

Agricultural production is a process combining land, the labor force, capital and knowledge. We propose the following production function for the agricultural sector

$$F_a = Z^q K_a^a N_a^b L_a^c, \quad a, b, c > 0, \quad a + b + c = 1, \quad q \geq 0, \tag{6.1}$$

in which Z is the knowledge stock, and K_a, N_a and L_a are the capital stock, the labor force and the land employed in the agricultural sector, respectively. Maximizing the profit by the agricultural sector yields the following conditions

$$r = aTF_a/K_a, \quad w = bTF_a/N_a, \quad R = cTF_a/L_a, \tag{6.2}$$

in which T = 1 - the tax rate levied upon the product. We assume that the government only obtains its income for supporting the university from the producers. There is no tax upon property and wage incomes. Moreover, the tax rate is identical for the two production sectors. We can relax this assumption by introducing heterogeneous tax rates upon production taxing property and wage incomes within the framework of this study.

Let C_a be the consumption of agricultural commodities by the population. Since we assume that agricultural goods cannot be saved, we have

$$C_a = F_a. \tag{6.3}$$

industry

Industrial production is carried out by combining knowledge, capital and the labor force. We specify the production function in the following form

$$F_i = Z^m K_i^\alpha N_i^\beta, \quad \alpha, \beta > 0, \quad \alpha + \beta = 1, \quad m \geq 0, \tag{6.4}$$

in which K_i and N_i are the capital stock and the labor force employed in the industrial sector, respectively. The marginal conditions are given by

$$r = \alpha p T F_i / K_i, \quad w = \beta p T F_i / N_i. \tag{6.5}$$

We now discuss the capital accumulation dynamics. The gross income, $Y(t)$, is given by

$$Y(t) = rK + wN + RL. \tag{6.6}$$

Let s $(0 < s < 1)$ denote the savings rate of the population from its current gross income. For simplicity, we assume that the savings rate is constant during the study period. The national savings is given by: sY. Capital accumulation is given by

$$dK/dt = sY/p - \delta K, \tag{6.7}$$

in which δ is the given depreciation rate of capital, K the total capital stock of the society.

Denote by C_i the consumption of industrial goods by the population. Since industrial production consists of investment and consumption, we have

$$C_i = F_i - sY/p. \qquad (6.8)$$

consumers' behavior

In this study, we do not distinguish between the behavior of consumers. For simplicity of analysis, we assume that consumption behavior is identical for the whole population. There is a "standard consumer" whose utility function represents the population. It is assumed that utility is a function of consumption levels of agricultural good, industrial goods and housing. For simplicity, we use the land utilized by households as a measurement of housing conditions. We specify the following utility function

$$U = C_a^u C_i^v L_c^y, \quad u, v, y > 0,$$

where L_c is the land distributed to the housing market. The total consumption budget is given by: $(1-s)Y$. The consumer problem is thus defined by

$$\max U, \quad \text{subject to: } C_a + pC_i + RL_c = (1-s)Y. \qquad (6.9)$$

The unique optimal solution is given by

$$C_a = us_0 Y, \quad C_i = vs_0 Y/p, \quad L_c = ys_0 Y/R, \qquad (6.10)$$

in which $s_0 = (1-s)/(u+v+y)$.

knowledge growth

We now specify the dynamics of knowledge. Changes in education systems, greater freedom of communications among people and nations, as well as other social and economic conditions may increase the level of knowledge. We consider three sources of knowledge accumulation. We assume that knowledge accumulation is positively related to scales of agricultural and industrial activity. We propose the following possible dynamics of knowledge

$$dZ/dt = \tau_a F_a/Z^c + \tau_i F_i/Z^\pi + \tau_p Z^g K_p^z N_p^o - dZ, \qquad (6.11)$$

in which K_p is capital, N_p denotes the scientists employed by the university, d is the fixed depreciation rate of knowledge, and τ_a, τ_i, τ_p are parameters. We call the terms, $\tau_a F_a/Z^c$, $\tau_i F_i/Z^\pi$, and $\tau_p Z^g K_p^z N_p^o$, the creativity of the agricultural sector, the industrial sector, and the university, respectively.

The term, $\tau_i F_i/Z^\pi$, measures the contribution of the industrial sector to knowledge growth. It implies that knowledge accumulation is positively related to production. The term, $1/Z^\pi$, implies that there are scale effects in knowledge accumulation. We can similarly interpret the term, $\tau_a F_a/Z^c$.

We now interpret $\tau_p Z^g K_p^z N_p^o$. This term implies that the creativity of the university is positively related to the number of scientists employed by the university and the size of its capital stock. On the one hand, as the knowledge stock is increased, the university may more effectively utilize traditional knowledge to discover new theorems. On the other hand, a large stock of knowledge may make the discovery of new knowledge difficult. This implies that the parameter, g, may be either positive or negative. We require: $-1 < g < 1$.

From the formation of knowledge accumulation and the production functions of the two sectors, we see that the economic system may exhibit increasing returns. The significance of introducing endogenous knowledge into growth theory has been discussed extensively by, e.g., Romer (1986, 1990), Grossman and Helpman (1991), Krugman (1990), Lucas (1988), and Zhang (1990, 1993a, 1993b).

the budget of the university

Since we assume that the financial resources of the university come from government taxes upon production, we have

$$(1-T)(F_a + pF_i) = wN_p + rK_p. \tag{6.12}$$

We now have to design a way to determine N_p and K_p. We assume that the government decides the number of scientists employed by the university and the level of its capital stock in the following way

$$N_p = nN, \quad K_p = kK, \quad 0 < n, k < 1, \tag{6.13}$$

in which n and k are the policy variables fixed by the government. We assume that n and k are exogenously given. We are interested in the effects of the policy parameters n and k upon the economic system. In this study, we assume that the tax rate is endogenously given in the sense that the government will tax the producers only to satisfy (6.12). That is, $(1 - T)$ is equal to $(wN_p + rK_p)/(F_a + pF_i)$ at any point of time.

population growth

Although there are a number of mathematical models which treat population growth as an endogenous process of economic development, only a few of them examine the possibility that knowledge may affect population growth (e.g., Becker, Murphy and Tamura, 1990, Becker and Barro, 1988).

In order to develop population dynamics, we would like to mention a few of the population growth models described in the literature. The simplest form of the "Malthusian growth model" assumes the following rule: $dN/dt = a_0N$, where $N(t)$ is the population at time t and a_0 is a constant. The logistic growth model, $dN/dt = a_0N(1 - b_0N)$, takes account of the checking effects of natural resources upon population growth. However, in relation to society, resources are not given, but rather "endogenous variables". To analyse how production affects population growth, Haavelmo (1954) suggests a model suitable for an agricultural economy: $dN/dt = a_0N(1 - b_0N/Y)$, where $Y(N)$ is the production of the society, dependent upon the labor input (with land fixed). Another formulation of population growth is given by: the population growth rate = $(\text{consumption})^v/(\text{wages})^u$, where $u > 0$ and $v > 0$. This formula can be interpreted as meaning that the demand for children will increase as current expenditure increase and decline as the wage rate increases. This is consistent with the microeconomic theory of fertility (Cigno, 1986, Becker and Barro, 1988). In a recent study of the interactions of human capital, population growth and economic growth, Becker, Murphy and Tamura (1990) assume that rates of return on investments in human capital rise rather than decline as the stock of human capital increases, at least until the stock becomes large. They show that if the stock of human capital is low, then people tend to have large families, and vice versa. Upon the basis of these studies, we propose the following population dynamics

$$dN/dt = VN\{(C/N)^\theta - h_1 Z^h (K/N)^j\}, \tag{6.14}$$

in which V is a positive adjustment speed parameter, θ ($0 \leq \theta < 1$), h_1 and j ($0 \leq j < 1$) are positive parameters.

The population growth rate, i.e., (dN/dt)/N is positively related to the current consumption level, and negatively related to human capital and capital per capita. It is not difficult to check that the population growth models mentioned above can be considered, in a broad sense, as a special case of (6.14).

full employment of the labor force, capital and land

The assumption that the labor force, capital and land are always fully employed yields the following equations

$$N_a + N_i = (1-n)N, \quad L_a + L_c = L, \quad K_a + K_i = (1-k)K, \tag{6.15}$$

in which L is fixed.

We have thus built the model which explains the distribution of the labor force, capital and land as well as dynamics of capital, knowledge, and the population. The system consists of 19 endogenous variables, K, K_a, K_i, N, N_a, N_i, L_a, L_c, Z, F_a, F_i, C_a, C_i, p, r, R, w, T, and Y, and of the same number of independent equations. We now show that the system has solutions under appropriate conditions.

6.3 EQUILIBRIA AND STABILITY

First, we show that the dynamics can be written in terms of three differential equations for K, Z and N. From (6.2) and (6.5), we have

$$N_a/N_i = (\alpha b/a\beta)K_a/K_i. \tag{6.16}$$

From (6.2), (6.3) and (6.10), we have: $L_c/L_a = y/ucT$, or

$$L_i = yL/(y + ucT), \quad L_a = ucTL/(y + ucT). \tag{6.17}$$

From (6.3), (6.6), (6.10), we have: $p = (1-ys_0-us_0)F_a/us_0F_i$. Utilizing $w = bTF_a/N_a = \beta pTF_i/N_i$, we get: $N_i/N_a = \beta(v + sy + su)/(1-s)bu$, or

$$N_i = \beta(v+sy+su)u_2N, \quad N_a = (1-s)buu_2N, \tag{6.18}$$

in which $u_2 = (1-n)/\{bu(1-s) + \beta(v+sy+su)\}$. Since $K_a + K_i = (1-k)K$ and $K_i/K_a = \alpha(v + sy + su)/a(1-s)u$, we have

$$K_a = a(1-s)uu_1K, \quad K_i = \alpha(v+sy+su)u_1K, \tag{6.19}$$

in which $u_1 = (1-k)/\{a(1-s)u + \alpha(v+sy+su)\}$.

From $p = (1-ys_0-us_0)F_a/us_0F_i$,

$$wN_p = nbNTF_a/N_a = nTF_a/(1-s)uu_2,$$

$$rK_p = akTF_aK/K_a = kTF_a/(1-s)uu_1,$$

we determine the tax rate, (1 - T), by

$$1 - T = (wN_p + rK_p)/(F_a + pF_i)$$

145

$$= Ts_0(n/u_1 + k/u_1)/(1-s)(1-ys_0)$$

i.e.,

$$1/T = 1 + s_0(n/u_2 + k/u_1)/(1-s)(1-ys_0).$$

This implies that the tax rate is a constant determined by the saving rate, government's research policy, the production structure and the preferences of the households.

Utilizing the above equations, we can express the dynamics in terms of K, Z and N as

$$dK/dt = s_1 f_i(K,Z,N) - \delta K,$$

$$dZ/dt = \tau_a f_a(K,Z,N) + r_i f_i(K,Z,N)/Z^\tau + \tau_p n_0 Z^g K^z N^\sigma - dZ,$$

$$dN/dt = VN\{(C/N)^\theta - h_1 Z^h (K/N)^j\}, \tag{6.20}$$

in which $n_0 = k^p n^\sigma$, $s_1 = s/(1-ys_0-us_0)$, $0 < s_1 < 1$,

$$f_i(K,Z,N) = m_1 K^\alpha Z^m N^\beta, \quad f_a(K,Z) = m_2 K^a Z^q N^b/Z^\prime,$$

$$C = (1-s)Y = (1-s)F_a/us_0 = m_0 K^a Z^q N^b,$$

$$m_1 = \beta^\beta \alpha^\alpha (v+sy+su) u_1^\alpha u_2^\beta > 0, \quad m_0 = (1-s)m_2/us_0$$
$$\tag{6.21}$$
$$m_2 = \{a(1-s)uu_1\}^a \{(1-s)buu_2\}^b L_a^c > 0.$$

We have established the economic model describing time-dependent paths of capital, knowledge and the population.

An economic equilibrium is determined as a solution of the following equations

$$s_1 f_i = \delta K, \quad \tau_a f_a + r_i f_i/Z^\tau + \tau_p n_0 Z^g K^z N^\sigma = dZ,$$
$$\tag{6.22}$$
$$(m_0 K^a Z^q N^{b-1})^\theta = h_1 Z^h (K/N)^j.$$

From the first and the last equations in (6.22), we have

$$K = A_1 Z^{m/\beta + \Lambda}, \quad N = A_0 Z^\Lambda, \tag{6.23}$$

in which $A_0 = \{(m_0^\theta/h_1)(s_1 m_1/\delta)^{(j-a\theta)/\beta}\}^{1/c\theta} > 0$, $A_1 = (s_1 m_1/\delta)^{1/\beta} A_0 > 0$, $\Lambda = (q\theta - h - jm/\beta + a\theta m/\beta)/c\theta$. Substituting (6.23) into the second equation in (6.22) yields

$$H(Z) = \Phi_1(Z) + \Phi_2(Z) + \Phi_3(Z) - d = 0, \tag{6.24}$$

in which

$$\Phi_1 = \tau_a^* Z^{x_1}, \quad \Phi_2 = \tau_i^* Z^{x_2}, \quad \Phi_3 = \tau_p^* Z^{x_3}, \tag{6.25}$$

where $\tau_a^* = \tau_a m_2 A_0^b A_1^{a^*} > 0$, $\tau_i^* = \tau_i m_1 A_0^\delta A_1^\alpha > 0$, $\tau_p^* = \tau_p n_0 A_0^\sigma A_1^z > 0$, $x_1 = q - \epsilon + (a+b)\Lambda + am/\beta - 1$, $x_2 = m - \pi + \alpha m/\beta + \Lambda - 1$, $x_3 = g + (\sigma+z)\Lambda + zm/\beta - 1$. We now try to determine the conditions under which $H(Z) = 0$ has solutions.

We exclude the case of $x_i = 0$, $i = 1, 2, 3$. One can directly check that if $x_i \geq 0$ for all i (i.e., $H(0) <$, $H(\infty) > 0$ and $H' > 0$ for $Z > 0$) or $x_i \leq 0$ for all i (i.e., $H(0) > 0$, $H(\infty) < 0$ and $H' < 0$ for $Z > 0$), then the system has a unique positive equilibrium. In any case of the remaining six combinations of $x_i > 0$ or $x_i < 0$, $i = 1, 2, 3$, we only prove the case of $x_1 > 0$, $x_2 < 0$, $x_3 < 0$. The other cases can be similarly checked. As $H(0) > 0$ and $H(\infty) > 0$, $H(Z) = 0$ has either no solution or multiple solutions. As

$$ZH'(Z) = x_1 \Phi_1(Z) + x_2 \Phi_2(Z) + x_3 \Phi_3(Z), \tag{6.26}$$

$H'(Z)$ may be either positive or negative, depending upon the parameter values. If $H(Z) = 0$ has more than two solutions, there are at least two values of Z such that $H'(Z) = 0$. Since $d(ZH')/dZ > 0$ strictly holds for $Z > 0$, it is impossible for $H'(Z) = 0$ to have more than one solution. Accordingly, $H(Z) = 0$ has either no solution or two solutions. A necessary and sufficient condition for the existence of two equilibria is that there exists a value of Z^* such that $H(Z^*) < 0$.

The Jacobian at an equilibrium is given by

$$J = \begin{pmatrix} -\beta\delta & m\delta K/Z & \beta\delta K/N \\ g_1 & g_2 & g_3 \\ g_4 & g_5 & g_6 \end{pmatrix} \qquad (6.27)$$

in which

$$g_1 = a\tau_a f_a/K + \alpha r_i f_i/KZ^\tau + z\tau_p n_0 Z^z K^{z-1} N^\sigma \geq 0,$$

$$g_2 = (q-1)_a a f_a/Z + (m-1) r_i f_i/Z^{\tau+1} + (g-1)\tau_a n_0 Z^z K^z N^\sigma/Z \leq 0, \qquad (6.28)$$

$$g_3 = \beta\tau_a f_a/N + \beta r_i f_i/Z^\tau N + \sigma\tau_p n_0 Z^z K^z N^{\sigma-1} \geq 0,$$

$$g_4 = VN(C/N)^\theta(a\theta-j)/K, \quad g_5 = VN(C/N)^\theta(q\theta-h)/Z$$

$$g_6 = V(C/N)^\theta(b\theta-\theta-j).$$

The three eigenvalues, ϕ_i ($i = 1, 2, 3$), are determined by

$$\phi^3 + D_1\phi^2 + D_2\phi + D_3 = 0, \qquad (6.29)$$

where $D_1 = \beta\delta - g_2 - V(C/N)^\theta(b\theta-\theta-j) > 0$, $D_3 = -|J|$,

$$D_2 = \begin{pmatrix} -\beta\delta & m\delta K/Z \\ g_1 & g_2 \end{pmatrix} + \begin{pmatrix} g_2 & g_3 \\ g_5 & g_6 \end{pmatrix} + \begin{pmatrix} \beta\delta & \beta\delta K/N \\ g_4 & g_6 \end{pmatrix}. \qquad (6.30)$$

The necessary and sufficient conditions for stability are known as the Routh-Hurwitz criterion: (i) $D_i > 0$; (ii) $D_1D_2 - D_3 > 0$. We can show that the system may be either stable or unstable, dependent upon the parameter values. In fact, as shown in the appendix, the system may be either stable or unstable even when the adjustment speed of the population is equal to zero, i.e., $V = 0$. As we are only concerned with the effects of changes in some important parameters, we will not discuss the stability conditions further.

In the remainder of the study, we assume that $x_i < 0$, i.e., $q + (a+b)A + am/\beta < \epsilon + 1$, $m + \alpha m/\beta + A < \tau + 1$, $g + (\sigma+z)A + zm/\beta < 1$. It is not difficult to see that, in a very broad sense, the assumption of $x_i < 0$ can be interpreted as meaning that the creativity of all the economic activities and the scientific research is not too high.

6.4　EFFECTS OF THE GOVERNMENT'S RESEARCH POLICY

The government may affect knowledge production in two ways. The first is to increase the number of scientists. The second is to improve material conditions such as instruments, buildings and books.

Taking derivatives of (6.23) and (6.24) with respect to n yields

$$\Phi dZ/dn = d_1\Phi_1 + d_2\Phi_2 + d_3\Phi_3, \tag{6.31}$$

in which $\Phi = -H' > 0$ (as $x_i < 0$) with H' defined by (6.26) and

$$d_1 = (1+b+a)\{cy/(y+ucT)\}dT/dn - (a+b)(j-a\Theta)/(1-n)c\Theta$$

$$- a(1-a)(c + b)/c(1-n) - b(1+b)/(1-n),$$

$$d_2 = -(1 + \alpha b/c + b\beta)/(1-n) + \{cy/(y+ucT)\}dT/dn$$

$$- \alpha(j-a\Theta)/(1-n)c\Theta,$$

$$d_3 = \sigma/n - \{z + zb + \sigma b/c\}/(1-n) + (\sigma+z)\{y/(y+ucT)\}dT/dn$$

$$- (\sigma+z)(j-a\Theta)/(1-n)c\Theta,$$

$$dT/dn = -T^2 s_0/s_n(1-s)(1-ys_0) < 0. \tag{6.32}$$

For simplicity of discussion, let $j - a\Theta \geq 0$, which implies that the checking force of wealth per capita upon population growth is stronger than that of the consumption per capita. This requirement, $j - a\Theta \geq 0$, guarantees: $d_1 < 0$ and $d_2 < 0$.

From (6.31), we see that if d_3 is negative, the level of knowledge will certainly be reduced as the nation increases its number of scientists. If d_3 is positive, the level of knowledge will either be reduced or increased, dependent upon the actual situations of the system. In d_3, only one term, σ/n, is positive. Accordingly, only when σ/n is appropriately large, may d_3 become positive. This implies that if the percentage of scientists in the population is low and the creativity of the scientists is high, then d_3 tends to be positive.

Because Φ_1, Φ_2 and Φ_3 represent the creativity in agriculture, industry, and the university, respectively, we may interpret (6.31) as follows: when $d_1 < 0$, $d_2 < 0$ and $d_3 > 0$, if the level of creativity in agriculture and the level of creativity in industry are extremely low, i.e., Φ_1 and Φ_2 being small, and the level of creativity of the university is high, i.e., Φ_3 being large, then an increase in the number of scientists will increase the knowledge stock of the nation in the long term.

Since scientific research requires national resources, it is acceptable to conclude that the effects of the number of scientists upon the knowledge stock of the nation are not only related to the creativity of scientists, but also to the creativity of farmers and workers. When the creativity of the farmers and workers is extremely limited and the number of scientists is very small, we may expect the level of knowledge to be increased if the nation increases its number of scientists. But when a nation already has a large number of scientists and its industrial activities have a high level of creativity, an increase in the number of scientists in the university may reduce the level of knowledge in the society. Accordingly, whether or not the government should spend more money on research is very dependent upon the actual situation of the system under consideration.

The effects of changes in n upon K and N are given by

$$(c\Theta Z/K)dK/dn = - Z\Theta(b+c)/(1-n) + Zc\Theta\{y/(y+ucT)\}dT/dn -$$

$$Z(j-a\Theta)/(1-n) + (c\Theta m/\beta + q\Theta - h - jm/\beta + a\Theta m/\beta)dZ/dn,$$

$$(1/N)dN/dn = - b/c(1-n) + \{y/(y+ucT)\}dT/dn -$$

$$\beta(j-a\Theta)/(1-n)\beta c\Theta + \{(q\Theta - h - jm/\beta + a\Theta m/\beta)Z/c\Theta\}dZ/dn.$$

It is rather difficult to explicitly conclude the effects of the government's research policy upon capital accumulation and population growth in the long term.

Utilizing $L_a = ucTL/(y + ucT)$, $F_a = Z^q K_a^a N_a^b L_a^c$, $F_i = Z^m K_i^\alpha N_i^\beta$, $r = aTF_a/K_a$, $w = bTF_a/N_a$, $R = cTF_a/L_a$, $p = (1-ys_0-us_0)F_a/us_0 F_i$, $Y = F_a/us_0$, one can directly check the effects of the government's policy upon land distribution, the outputs of the agricultural and industrial sectors, the interest rate, the wage rate, the land rent, prices and the national income. Since the conditions are difficult to interpret explicitly, we will not represent them here.

As one can similarly carry out the comparative statics analysis with respect to a shift in the parameter, k, we omit the analysis.

6.5 KNOWLEDGE ACCUMULATION EFFICIENCY, τ_a, τ_i AND τ_p

Since τ_a, τ_i and τ_p have similar effects upon the system, it is sufficient for us to examine the impact of changes in τ_p on the system.

Taking derivatives of (6.22) and (6.23) with respect to τ_p yields

$$\Phi dZ/d\tau_p = \Phi_3/\tau_p > 0, \quad (Z/K)dK/d\tau_p = (m/\beta + A)dZ/d\tau_p, \quad (6.33)$$

$$(N/Z)dN/d\tau_p = AdN/d\tau_p.$$

Accordingly, an improvement in the efficiency of scientists' research will certainly increase the knowledge stock in the long term. From $A = (q\Theta - h - jm/\beta + a\Theta m/\beta)/c\Theta$, we see that $dN/d\tau_p > 0$ only when $q + am/\beta > h/\Theta + jm/\Theta\beta$. That is, as the efficiency of scientists' research is improved, the population will be increased only when the utilization efficiency of knowledge of the agricultural and industrial sectors (i.e., q and m, respectively), the marginal product of capital of the agricultural and industrial sectors (i.e., a and α, respectively), and the checking force of consumption upon population growth (i.e., Θ) are appropriately high, but the checking force of knowledge and wealth upon the population growth (i.e., h and j, respectively) are relatively low. We may conclude that $dN/d\tau_p$ tends to be negative in a society in which knowledge and wealth have strong (negative) effects and consumption has weak effects upon population growth. As $m/\beta + A = (q\Theta + (\Theta-b\Theta-j)m/\beta - h)/c\Theta$, we see that $dK/d\tau_p$ may be either positive or negative.

For simplicity of discussion, in the remainder of the section let $A = 0$, i.e., $dN/d\tau_p = 0$, $dZ/d\tau_p > 0$ and $dK/d\tau_p > 0$. We directly have

$$dF_a/d\tau_p > 0, \; dF_i/d\tau_p > 0, \; dR/d\tau_p > 0, \; dw/d\tau_p > 0, \; dY/d\tau_p > 0.$$

As the efficiency of scientists' activities is improved, the output of the two economic sectors, the land rent, the net income, and the wage rate tend to be increased in the long term. The impact upon the interest rate and the price of the industrial good is given by

$$(Z/r)dr/d\tau_p = \{q - (1-a)m/\beta\}dZ/d\tau_p, \quad (6.34)$$

$$(Z/p)dp/d\tau_p = \{q - (1-a)m/\beta\}dZ/d\tau_p.$$

It is important to note that $dr/d\tau_p$ and $dp/d\tau_p$ has the same sign. That is, if the interest rate declines due to the increased creativity of scientists, the price of the industrial good will also decline, and vice versa. Remembering that q and m represent the marginal product of knowledge of the agricultural and industrial sectors, respectively, and a and α (= 1 - ß) represent the marginal product of capital of the agricultural and industrial sectors, respectively, we can interpret the economic implications of the conditions for $dr/d\tau_p$ and dp/τ_p to be negative.

6.6 EFFECTS OF THE POPULATION ADJUSTMENT PARAMETER, h_1

The definition of h_1 in (6.14) implies that h_1 is determined by the relative strength of checking forces, $(C/N)^\theta$ and $Z^h(K/N)^j$, upon population growth. An increase in h_1 may be interpreted as meaning that the checking force of wealth and knowledge upon population growth becomes more significant in comparison to that of the consumption level.

The impact of h_1 upon K, Z and N is given by

$$\Phi dZ/dh_1 = - [(a+b)\Phi_1 + \Phi_2 + (\sigma+z)\Phi_3]/c\Theta h_1 < 0,$$

$$(Z/K)dK/dh_1 = (m/ß + A)dZ/dh_1 - Z/c\Theta h_1, \qquad (6.35)$$

$$(Z/N)dN/dh_1 = AdZ/dh_1 - Z/c\Theta h_1.$$

If A is positive, $dZ/dh_1 < 0$, $dK/dh_1 < 0$ and $dN/dh_1 < 0$; if $A < 0$, $dZ/dh_1 < 0$, and dK/dh_1 and dN/dh_1 may be either positive or negative.

It is easy to provide analytical results about the effects of changes in h_1 upon the output of the agricultural sector and the industrial sector, net income, the price of industrial commodity, interest rate, wage rate, and land rent.

6.7 THE COMPLEXITY OF THE EFFECTS OF SAVING BEHAVIOUR

Different economic theories, such as Keynesian and neo-classical economics, have various points of view about the effects of the savings rate upon economic development. One can also hardly get a definite conclusion between the savings rate and economic development from empirical studies (see, e.g., Chenery and Srinivasan, 1988, 1989). This section provides some analytical results to illustrate the complexity of the impact of the savings rate upon economic development.

The effects of changes in s upon knowledge, capital and the population are given by

$$\Phi dZ/ds = (m_2^* + bA_0^* + aA_1^*)\Phi_1 + (m_1^* + \beta A_0^* + \alpha A_1^*)\Phi_2$$

$$+ (\sigma A_0^* + zA_1^*)\Phi_3$$

$$(1/K)dK/ds = (1/A_1)dA_1/ds + (m/\beta Z + A/Z)dZ/ds,$$

$$(1/N)dN/ds = (1/A_0)dA_0/ds + (A/Z)dZ/ds, \qquad (6.36)$$

in which

$$m_1^* \equiv (1/m_1)dm_1/ds = (\alpha/u_1)du_1/ds + (\beta/u_2)du_2/ds,$$

$$m_2^* \equiv (1/m_2)dm_2/ds = -(a+b)/(1-s) + (a/u_1)du_1/ds$$

$$+ (b/u_2)du_2/ds + \{cy/(y+ucT)\}dT/ds,$$

$$A_0^* \equiv (1/A_0)dA_0/ds = (1/c\Theta m_2)dm_2/ds +$$

$$(j-a\Theta)(1-s_1)/s\beta c\Theta(1-s) + \{(j-a\Theta)/m_1 c\Theta\}dm_1/ds,$$

$$A_1^* \equiv (1/A_1)dA_1/ds = (1/A_0)dA_0/ds + (1-s_1)/\beta s(1-s)$$

$$+ (1/\beta m_1)dm_1/ds,$$

where

$$(1/u_2)du_2/ds = (bu - \beta y - \beta u)u_2/(1-n),$$

$$(1/u_1)du_1/ds = (au - \alpha y - \alpha u)u_1/(1-k),$$

$$dT/ds = s_0 T^2[ys_0(n/u_2 + k/u_1)/(1-s)(1-ys_0) + n(bu - \beta y$$

$$- \beta u)/(1-n) + k(au - \alpha y - \alpha u)/(1-k)]/(1-s)(1-ys_0).$$

Obviously, one can hardly conclude what effects the savings rate have upon the system, even in simple cases.

6.8 CONCLUDING REMARKS

This study suggested a dynamic model to describe an economic structure with endogenous capital, knowledge and population. We found that the dynamic system may have either unique or multiple equilibria, dependent upon the economic structure, the creativity of different activities and the population adjustment characteristics.

The study is based upon many other assumptions. Some of them can be easily relaxed. For instance, it is reasonable to consider the savings rate as a function of knowledge, capital, income and the interest rate. We may assume that the unemployment rate is a function of the production conditions and thus we can treat employment as an endogenous variable. It is also important to examine the impact of other policy variables - such as tax cuts in different sectors upon the economic structure. Indeed, the model can be extended to the case of multiple regions along the lines of multiple regional modelling (e.g., Allen and Sanglier, 1979, Batten, Kobayashi and Andersson, 1989, Nijkamp, 1986, Zhang, 1991a, 1991b, 1992a).

APPENDIX: STABILITY WITH THE CONSTANT POPULATION

As the stability conditions of the three dimensional system are so difficult to discuss, we simplify the problem by assuming $V = 0$, i.e., N is constant. In this case, an equilibrium is determined by

$$s_i f_i = \delta K, \quad \tau_a f_a + r_i f_i / Z^\pi + \tau_p n_0 Z^g K^z N^\sigma = dZ.$$

From $s_i f_i = \delta K$, we have: $K = A_1 Z^{m/\beta}$ in which $A_1 = N(s_1 m_1/\delta)^{1/\beta} > 0$. An equilibrium value of Z is also determined by (6.24) with τ_a^*, τ_i^* and τ_p^* being positive constants, and $x_1 = q - \epsilon + am/\beta - 1$, $x_2 = m - \pi + \alpha m/\beta - 1$, $x_3 = g + zm/\beta - 1$. The conditions for the existence of equilibria are the same as in the case of V being positive. One can directly check the stability conditions. One can easily prove the following proposition.

Proposition

We omit the case of $x_1 = x_2 = x_3 = 0$.

If $x_i \leq 0$ for all i, the system has a unique stable equilibrium.

If $x_i \geq 0$ for all i, the system has a unique unstable equilibrium.

In each of the remaining six combinations of $0 > x_i$ or $x_i < 0$ for all i, the system has either no equilibrium or two equilibria. When the system has two equilibria, the one with higher values of K and Z is unstable and the other one is stable.

ACKNOWLEDGEMENT

I am grateful for the comments from the participants in the International Workshop on Nonlinear Evolution of Spatial Economic Systems.

REFERENCES

Allen, P.M. and M. Sanglier, 1979, Dynamic Model for Growth in a Central Place System. **Geographical Analysis**, 11, pp. 256-272.

Batten, D.F., Kobayshi, K. and Å.E. Andersson, 1989, Knowledge, Nodes and Networks: An Analytical Perspective, (Å.E. Andersson, D.F. Batten and C. Karlsson, eds.), **Knowledge and Industrial Organization**, Springer-Verlag, Berlin.

Becker, G.S. and R.J. Barro, 1988, A Reformulation of the Economic Theory of Fertility. **Quarterly Journal of Economics**, 103, pp. 1-25.

Becker, G.S., Murphy, K.M., and R. Tamura, 1990, Human Capital, Fertility, and Economic Growth. **Journal of Political Economy**, 98, pp. 12-37.

Chenery, H. and T.N. Srinivasan, (ed.), 1988, **Handbook of Development Economics**, Vol.I, North-Holland, Amsterdam.

Chenery, H. and T.N. Srinivasan, (ed.), 1989, **Handbook of Development Economics**, Vol.II, North-Holland, Amsterdam.

Cigno, A., 1986, Fertility and the Tax-Benefit System, **Economic Journal**, 96, pp. 1035-1051.

Dendrinos, D.S. and M. Sonis, 1990, **Chaos and Socio-Spatial Dynamics**, Springer-Verlag, New York.

Grossman, G.M. and E. Helpman, 1991, **Innovation and Growth in the Global Economy**, The MIT Press, Massachusetts.

Haavelmo, T., 1954, **A Study in the Theory of Economic Evolution**, North-Holland, Amsterdam.

Haken, H., 1977, **Synergetics: An Introduction**, Springer-Verlag, Berlin.

Haken, H., 1983, **Advanced Synergetics**, Springer-Verlag, Berlin.

Krugman, P.R., 1990, **Rethinking International Trade**, The MIT Press, Cambridge, Massachusetts.

Lucas, R.E., 1988, On The Mechanics of Economic Development, **Journal of Monetary Economics**, 22, pp. 3-42.

Nicolis, G. and I. Prigogine, 1977, **Self Organization in Nonequilibrium Systems**, Wiley, New York.

Nijkamp, P., (ed.), 1986, **Technological Change, Employment and Spatial Dynamics**, Springer-Verlag, Berlin.

Romer, P.M., 1986, Increasing Returns and Long-Run Growth, **Journal of Political Economy**, 94, pp. 1002-1037.

Romer, P.M., 1990, Endogenous Technological Change, **Journal of Political Economy**, 98, no.5, pp. 71-102.

Rosser, J.B. Jr., 1991, **From Catastrophe to Chaos: A General Theory of Economic Discontinuities**, Kluwer Academic Publihers, Boston.

Weidlich, W. and G. Haag, 1983, **Quantitative Sociology**, Springer-Verlag, Berlin.

Wilson, A.G., 1981, **Catastrophe Theory and Bifurcation: Application to Urban and Regional Systems**, Croom Helm, London.

Zhang, W.B., 1990, **Economic Dynamics - Growth and Development**, Springer-Verlag, Berlin.

Zhang, W.B., 1991, **Synergetic Economics - Time and Change in Nonlinear Economics**, Springer-Verlag, Berlin.

Zhang, W.B., 1991a, Regional Dynamics with Creativity and Knowledge Diffusion, **The Annals of Regional Science**, 25, pp. 179-191.

Zhang, W.B., 1991b, Economic Development with Creativity and Knowledge Diffusion, **Socio-Spatial Dynamics**, 2, pp. 1-12.

Zhang, W.B., 1992, A Development Model of Developing Economies with Capital and Knowledge Accumulation, **Journal of Economics**, 55, pp. 43-63.

Zhang, W.B., 1992a, Trade and World Economic Growth - Differences in Knowledge Utilization and Creativity, **Economic Letters**, 39, pp. 199-206.

Zhang, W.B., 1993, An Urban Pattern Dynamics with Capital and Knowledge Accumulation, **Environment and Planning A**, 25.

Zhang, W.B., 1993, Location Choice and Land Use in an Isolated State - Endogenous Capital and Knowledge Accumulation, **The Annals of Regional Science**, 27.

PART B: SPATIAL REPRESENTATIONS OF NONLINEAR DYNAMIC SYSTEMS

CHAPTER 7
MICROECONOMICS AND THE DYNAMIC MODELLING OF SPATIAL SYSTEMS
W.D. Macmillan

7.1 INTRODUCTION

The articulation of new methods for analyzing the dynamics of spatial systems is a welcome development. Of all the weaknesses in conventional methods, the inadequacy of the treatment of time has been the most glaring. The adoption of the mathematics of nonlinearity has helped us to address some old problems in novel ways and it has opened up some important new lines of inquiry. However, some of the ways in which nonlinear dynamics has been incorporated into spatial analysis has, in my view, failed to resolve the key problems associated with producing satisfactory theories of spatial dynamics and may even have exacerbated them. I will argue that older approaches, which seem to have been largely forgotten, provide better stems than most of the currently dominant approaches on which to graft the methods of nonlinear mathematics.. In particular, I will advocate the redevelopment of ideas associated with recursive programming. They appear to offer the opportunity to elaborate new theories without the restrictions attendant on the use of what I will refer to as the planning-model strategy.

7.2 MORPHOLOGICAL IMPERIALISM

The swirl of ideas in and around disciplines has its own, undoubtedly nonlinear, dynamics. The ecological analogy is hard to resist. The island of the spatial sciences is constantly invaded by new species of ideas. Some become dominant, others find a small niche and coexist with earlier arrivals, and many fail to establish themselves at all. Harris (1985) has an alternative picture; he describes the expansions associated with particular model forms and modelling strategies as 'morphological imperialism', echoing the idea of cultural imperialism. These analogies invite us to consider the extent to which various modelling species or cultures have come to dominate the spatial sciences. They also invite us to consider whether each new model form should be treated as an unwelcome invader, to be repulsed when the opportunity arises, or assimilated as a useful addition to our methodological stock. If assimilation appears to be desirable, the analogies suggest that we should go on to examine how it should be achieved: should we rely on fusion with the currently dominant species or culture, or promote a partnership with a well-matched but less-dominant model form?

The likely success of a new species or culture depends on the fertility of the invaded territory and the indigenous competition. The spatial sciences have an evident need for dynamic analysis and have lacked a thriving dynamic modelling school, so the conditions are certainly conducive to colonization. The dearth of dynamic modelling is somewhat surprising. Despite having drawn extensively from economics over the years, spatial analysts have made few attempts to develop spatial versions of theories of growth and business cycles - arguably the major areas of dynamic inquiry within economics. Indeed, the employment of macro-economic ideas in general has been remarkably restricted in a spatial context.

Microeconomics, on the other hand, has been a great inspiration, providing the foundations both for the various branches of location theory and for spatial versions of general equilibrium analysis. This micro-economic, location-theoretic approach can be thought of as one of the two great traditions of spatial analysis - one of the two major species currently flourishing on the spatial-analysis island. The other is the more instrumentalistic planning-model tradition, which I take to include most spatial interaction and land use-transportation models. This is arguably the dominant species.

These two traditions are, of course, closely interrelated but they still have distinctive features. In particular, they differ both in the formality and consistency with which they treat economic agency, equilibrium and disequilibrium. It will be argued below that this is of considerable significance for the successful development of dynamic theories of spatial systems.

Most nonlinear dynamic models within the spatial sciences belong to the planning-model tradition. This applies to the spatial-interaction work stemming from Wilson's writing on catastrophe theory and the associated landmark paper by Harris & Wilson (1978), and it applies to the applications of chaos theory by Reggiani (1990) and Nijkamp & Reggiani (1988). The earlier foray by Forrester (1969) into urban dynamics, the more sophisticated use of ecological dynamics by Allen & Sanglier (1981), the related work of Dendrinos & Mullally (1985), and the applications by Batty and others of fractal geometry (see, for example, Batty, Longley, & Fotheringham (1989)) can be categorized in the same way. None of these approaches pays much formal attention to microeconomic behaviour. The essence of the argument in this paper is that satisfactory development of dynamic spatial theory requires the resolution of certain problems within the microeconomic, location-theoretic school and that these problems will remain unresolved unless we adopt a different - or, at least, broader - approach to the introduction of nonlinear dynamics. To put it another way, the morphological imperialism *within* the spatial sciences has placed the planning-model tradition in an advantageous position to exploit nonlinear dynamics but it is the other tradition which we should concentrate on to achieve further advances.

7.3 THE DEVELOPMENT OF THE MICROECONOMIC, LOCATION-THEORETIC SCHOOL

The microeconomic, location-theoretic school is founded on the classical theories of location and land use (LT in Figure 1) from von Thünen, through Weber to Lösch. Developments in computing and optimization theory in the sixties and seventies led to attempts to formalize, generalize and operationalize the classics in a discrete-space, mathematical programming (MP) form. Operationalization was, arguably, the dominant motive of those involved in this work. Economists with primarily theoretical interests continued to work in the continuous-space tradition on the nature of monocentric cities, creating the new urban economics (NUE).

The use of MP models allowed the classics to be cast into a non-rigorous, partial equilibrium form. MP models did two things: they acted as a set of pseudo-axioms and as a structure for computing (or simulating) the behaviour of the axiomatized systems. In theoretical terms, this approach had two clear deficiencies. The existence of equilibrium was presumed rather than proved and this presumption was often ill-founded. Second, the nature of the microeconomic assumptions implied by (or claimed for) the models was often unclear. A partial resolution of both of these problems was achieved via the embedding of various programming models within general equilibrium theory. The work of Takayama and Judge (1971), for example, can be viewed as consisting of prototype computable general equilibrium (CGE) models, although they were not referred to as such at the time.

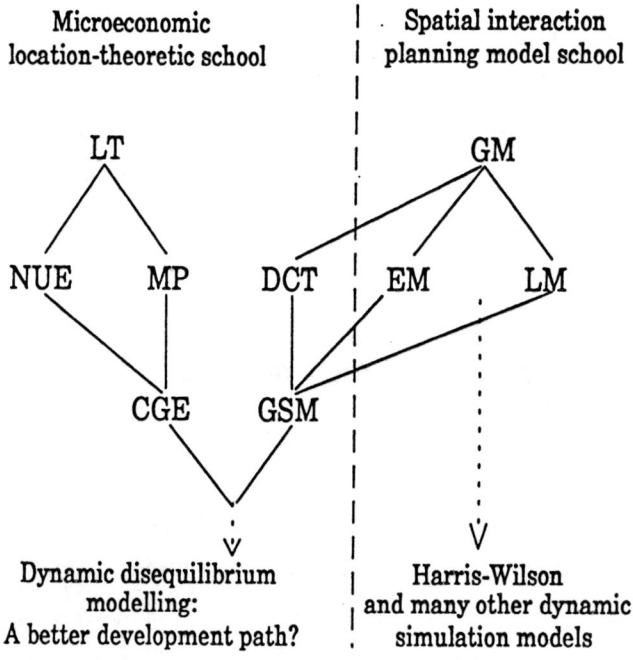

Figure 1 A highly-simplified schematic representation of the development of the two modelling schools (see the text for notation). In microeconomic terms, the fields on the left of the dashed line are relatively well-founded, whereas those on the right are, at best, problematic. Many (arguably most) modern dynamic simulation models belong on the right hand side of the figure. However, the tradition represented by the left hand side may prove more fertile in the long run.

Meanwhile, work in NUE progressed from monocentric, continuous-space models of cities to discrete-space general equilibrium simulation using CGE models. Mills (1974) book set the standard. The adoption of numerical simulation was necessitated by the analytical intractability of key questions concerning urban equilibria. It involved the division of space into discrete areas, typically annuli around a single centre rather than a more general set of zones. Nevertheless, this brought about a measure of convergence between the NUE and MP approaches.

7.4 THE DEVELOPMENT OF THE SPATIAL-INTERACTION, PLANNING-MODEL SCHOOL

The gravity model (GM in Figure 1) can be thought of as the forerunner of the work in the spatial-interaction, planning-model school. Gravity modelling was weaker theoretically than the microeconomic approach but stronger empirically. For transport planners in particular, empirical strength was crucially important. They took the view that "the application of such economic models as linear programming was an exercise in futility" (Harris (1985)).

The Lowry Model (LM) had much the same strength and weakness. For a start, it was essentially descriptive. According to Anas (1987), "the [Lowry] model is not economic but physicalistic in nature: it allocates physical quantities such as jobs, households etc. without any regard to the prices which resolve such an allocation process in real markets". What is more, "urban modellers were by and large content to write computer programs which were presented as black boxes. As long as these tools could produce and print out certain forecasts they were considered immune from any scrutiny into their mathematical form." This theoretical weakness manifested itself both in the absence of microeconomic support and in the presumption that equilibria exist. Yet "in possibly many cases the Lowry model does not have a solution. Apparently, many scholars who worked on extending the Lowry model are unaware of this problem" (Anas op.cit.).

The developments that have flowed from gravity and Lowry modelling are, of course, many and various. In particular, entropy maximization (EM) gave spatial interaction work a similar formalism to that of the MP microeconomic school. Indeed, an early interpretation of entropy maximization models was that they described sub-optimal behaviour on the grounds that their spatial interaction patterns tended to that of an equivalent linear programming problem as the transport cost tended to its minimum value.

Later, the development of discrete choice theory (DCT) provided a microeconomic rationale for spatial interaction systems. Williams (1977) group surplus maximization (GSM) models then provided the basis for unifying the entropy approach, the Lowry model, and discrete choice theory and, simultaneously, gave the spatial-interaction school the same welfare maximising, mathematical programming form as the microeconomic, location-theoretic school.

7.5   COMMON GROUND AND COMMON PROBLEMS

Further development of these ideas (see, in particular Anas (1984)) led to the cross-fertilization of the group surplus maximization and computable general equilibrium approaches. As a result, the first fruits of a unified approach to spatial economic analysis have started to appear.

Impressive though this development has been, it leaves several major problems unresolved. The first of these concerns the nature of equilibria in spatial systems. Typically, they are instant landscapes formed from current resources and conditions. New landscapes result from new resources and conditions. In other words, there tends to be no consideration of time beyond the comparative static. The second problem has to do with the existence of equilibria. Although some location theoretic models can be embedded within general equilibrium theory, some cannot without significant and damaging modification. This problem is particularly noticeable in the context of central place theory and has been demonstrated fairly formally (and famously) for a class of interacting industrial locators by Koopmans & Beckmann (1957).

The next problem centres on the stability of equilibria. Stability analysis is rarely used in spatial studies, despite the significance and familiarity of the Hotelling problem. Even if it could be established that certain central place formulations and certain variants of Koopmans-Beckmann systems are consistent with the existence of equilibria, they may be ephemeral. Next, there are real difficulties about the computation of equilibria. In an NUE context, King (1977) reported that "computation time rises roughly with n^4 where n is the number of commodities". Similar problems arise elsewhere. It is important to note that the difficulty of the computational task is a direct consequence of the instant-landscape, comparative-static conception of the modelling problem. Finally, and intimately connected with these other points is the implausibility of the micro-economic foundations of the common, general-equilibrium formulations. The resolution of all these problems is connected to the treatment of time. It is important, therefore, in choosing a dynamic modelling approach, to get it right.

7.6 NONLINEAR DYNAMICS IN THE SPATIAL-INTERACTION, PLANNING-MODEL TRADITION

The Harris & Wilson (1978) approach to modelling dynamics is particularly important in this debate because it has been elaborated to cover most of the branches of location and land use theory. It is an approach which has the ability to describe both catastrophic changes and chaotic behaviour. In terms of Figure 1, it resides firmly in the spatial-interaction, planning-model school and is characteristic of the school's strengths and weaknesses. The basic nature of the strategy stemming from the Harris & Wilson paper is as follows (this account is based on Clarke & Wilson (1985)).

Spatial interactions, T_{ij}, are a function of the amount of production activity at i, Z_i, the demand for the output of that activity at j, X_j, and the generalized travel cost, c_{ij}:

$$T_{ij} = T_{ij}(X_j, Z_i, c_{ij}) \qquad \forall ij \qquad (7.1)$$

The total value of the product of the activity at i is given by D_i, and the cost of this production activity is C_i, where

$$D_i = \sum_j T_{ij} \qquad \forall i \qquad (7.2)$$

$$C_i = C_i(Z_i) \qquad \forall i \qquad (7.3)$$

Dynamics is introduced by the assumption that

$$\frac{\partial Z_i}{\partial t} = \varepsilon(D_i - C_i) f(Z_i) \qquad \forall i$$

where t time, ε is a parameter and $f(Z_i)$ is some function of Z_i. The equilibrium condition associated with this equation is clearly

$$D_i = C_i \qquad \forall i$$

Substituting in the expressions for D_i and C_i and taking $f(Z_i)$ to be just Z_i, the differential equation becomes

$$\frac{\partial Z_i}{\partial t} = \varepsilon \left[\sum_j T_{ij}(X_j, Z_i, c_{ij}) - C_i(Z_i) \right] Z_i \qquad \forall i \qquad (7.4)$$

A more complex version of the model adds price adjustment mechanisms to the above activity adjustment process. The system of differential equations is then of the following form:

$$\frac{\partial Z_i}{\partial t} = \varepsilon_1 \left[\sum_j T_{ij}(X_j, Z_i, p_j, c_{ij}) - C_i(Z_i) \right] Z_i \qquad \forall i$$

$$\frac{\partial p_j}{\partial t} = \varepsilon_2 \left[\sum_i T_{ij}(X_j, Z_i, p_j, c_{ij}) - C_i(Z_i) \right] p_j \qquad \forall j$$

where p_j is the price associated with the activity at j, and ε_1, ε_2 and ε_3 are parameters.

These ideas have been operationalized in an agricultural land use context by Wilson & Birkin (1987). A modified version of their model is presented in Tables 1 and 2 alongside a modified version of the Day & Tinney model, which will be considered later. The two models have been altered slightly to accentuate their similarities and to facilitate comparison. One of the alterations is to present the Wilson & Birkin model in an optimization form. The solution to the optimization element of the problem is

$$Y_{ij}^k(t) = A_i^k(t) Z_i^k(t) W_j^k(t) \exp(-\beta c_{ij}^k)$$

(cf. equation (7.1)). The revenue and cost terms differ slightly from equations (7.2) and (7.3). They take the form

$$D_i^k(t) \equiv \sum_j \hat{p}_j^k(t) Y_{ij}^k(t)$$

$$C_i^k(t) \equiv \sum_j (c_{ij}^k + v_i^k) Y_{ij}^k(t)$$

For notational convenience, a net revenue term $E_i^k(t)$, is introduced, where

$$E_i^k(t) \equiv D_i^k(t) - C_i^k(t)$$

$$= \sum_j (\hat{p}_j^k(t) - c_{ij}^k - v_i^k) Y_{ij}^k(t)$$

The differential equation (7.4) is translated into the difference equation form

$$Z_i^k(t) - Z_i^k(t-1) = \sum_{m \neq k} \delta_i^{mk}(t-1) \varepsilon_i^m \frac{a_i^m}{a_i^k} Z_i^m(t-1)$$

where the value of δ_i^{mk} is determined by the relative sizes of the $E_i^k(t)$ terms for all j and k. Thus, the essential structure of the Harris & Wilson approach is maintained.

7.7   SOME SHORTCOMINGS OF DYNAMIC SPATIAL-INTERACTION MODELLING

Although this approach delivers a good deal in terms of the description of nonlinear dynamic behaviour, it has some significant shortcomings.

1) The nature of the dynamic process

The notion of a 'modified divergent' path to equilibrium is an important feature of the approach. It is used because "the basic state is unstable, but dampening the adjustment procedure allows stability to be achieved" (Wilson & Birkin (1987)). Parameters are adjusted to ensure convergence occurs (the ε terms). Thus, there is an artificial requirement that the process must tend to an equilibrium state. Also, the dynamics of this process are conceived in partly mechanical, as opposed to economic, terms.

2) Failure to distinguish between dynamic equilibrium and economic equilibrium

There is an assumption that a system of the type described is capable of existing in a dynamic equilibrium state when microeconomic theory casts doubt on the existence of an economic equilibrium state. In the case of the central place work, the theory is crucially dependent on increasing returns to scale in production but the model cost function implies a continuous linear technology. The model assumes equilibria will exist, but the theory gives no such assurance.

3) Inadequate microeconomic foundations

It is unclear whether or not any formal set of microeconomic assumptions will support the models. In the agriculture case, destination attraction is based on the money value of sales but there is no underpinning microeconomic rationale for this (and it cannot be borrowed from discrete choice theory in these circumstances). The rent term is present in the cost functions in some models but not others "since rent is fixed after the fact it is not an element of the cost function". In a realistic microeconomic description, rent would have to be paid, whether or not it is paid before or after 'the fact'. In reality, 'the fact' is an artefact of the algorithmic structure of the model. As noted earlier in connection with the Lowry model, this is a characteristic feature of the spatial-interaction, planning-model school.

4) Failure to address the problems of dynamic microeconomic behaviour.

The models fail to take the opportunity to explore how agents behave through time. This assumes great importance if the necessity to found operational models on sound theory is recognized. Moreover, sound theory may also lead to better computational procedures so there are strong operational arguments for examining alternative microeconomic foundations for dynamic spatial models.

7.8 MICROECONOMIC EQUILIBRIUM DYNAMICS

The standard, static, microeconomic approach has numerous deficiencies: no explicit treatment of time (so no treatment of market entry and exit), convexity of production and consumption possibilities, perfect knowledge, perfect rationality, perfect competition, and market clearing. The clear message of the spatial sciences is that all of these assumptions are untenable, at least at the level of spatial disaggregation at which we tend to work. Indeed, Marris has argued in a recent book on macroeconomic simulation that the notion of perfect competition is quite untenable even at the national level (Marris (1991)).

All of the above constructs, with the possible exception of the notion of set convexity, owe something to the collapse of time in the comparative static approach. There are numerous alternative approaches to the treatment of time, some of which are outlined below. Quite a number of these alternatives have found their way into the spatial sciences but none has established a significant niche.

The first is the idea of temporary equilibrium. This might be regarded as a modest extension of comparative statics, in which equilibrium is achieved in each time period over a number of successive periods.

Intertemporal equilibrium introduces time through a multiperiod extension of the static problem. That is, equilibrium is achieved between time periods over the time horizon of the model. This allows the treatment of intertemporal flows (storage), depreciation, and growth. However, there is an implicit assumption of perfect foresight so the control is open-loop. In other words, the optimal decision for all time periods can be made initially (see the quadratic programming, multi-period, multi-regional equilibrium models of Takayama & Judge (1971) for an example in the MP tradition).

A slightly more sophisticated variant would be to specify intertemporal equilibrium models with uncertainty, involving optimal stochastic control as opposed to optimal deterministic control. Mean-variance programming formulations could be used, extending the static, spatial, portfolio-theoretic models that have been applied to industry and agriculture by Cromley and Hanink (see, for example, Cromley (1982)). Dynamic programming formulations would be an alternative.

Rational expectations equilibrium is more sophisticated. It involves optimal adaptive control - learning about, and forecasting, system parameters (prices, say) - in a manner that produces self-fulfilling expectations. For this work, dynamic programming appears to be a necessity rather than an option. The difficulty with rational expectations is that, along with other equilibrium approaches, it is simply implausible because it requires perfect rationality and perfect competence.

Much more plausible and appealing is the bounded rationality articulated in the recursive programming models of Richard Day and others. Recursive programming models, which can incorporate both equilibrium and disequilibrum mechanisms, have been developed for the analysis of spatial problems both by Day and his collaborators and by Takayama and Judge. Perhaps the simplest application is the dynamic von Thünen model by Day & Tinney (1969).

7.9 RECURSIVE PROGRAMMING

A comparison between this model and that of Wilson & Birkin (1987) illustrates the difference between the location-theoretic and planning-model schools (see Tables 1 and 2 again). It also suggests some possible future developments.

The Day & Tinney model has a clear microeconomic foundation which allows local product specialization but no cross-hauling (Table 2, column 1, row 1); decision makers have to rely on imperfect information so they behave in a boundedly rational fashion (row 2); temporary equilibrium is achieved in the goods markets (row 3); and there is disequilibrium in the land markets, assuming that fixed 'overheads' can be treated as rents (row 4). The Wilson & Birkin model, on the other hand, exhibits cross-hauling -

one of the empirical strengths of spatial interaction work - but no specialization and the model has no identifiable microeconomic foundation (Table 2, column 2, row 1); the treatment of prices is equivalent to a microeconomic assumption of perfect foresight (row 2); product markets are in disequilibrium (row 3); and land markets can be thought of as being in temporary equilibrium since all residual income is siphoned off as rent (row 4).

I hope to show in a future paper that it is possible to construct a hybrid model from the above prototypes within the CGE/GSM paradigm and that such a model has all the advantages of the spatial interaction school without the disadvantages of weak microeconomic foundations. I also hope to demonstrate that recursive programming ideas open up a rich vein that we have hardly begun to tap. For example, the use of a survival condition in recursive programming models leads to the formalization of the idea of homeostatic behaviour, which may well be more appropriate than optimising behaviour as a descriptor of certain classes of agents. In particular, modifications of this model have potential applications in the context of poor spatial economies where agents are faced with the possibility of entitlement failure. The general notions of bounded rationality and recursive programming have much wider applications (see Day & Groves (1975)).

7.10 MICROECONOMIC DISEQUILIBRIUM DYNAMICS

Both of the models in Tables 1 and 2 have economic equilibrium and disequilibrium elements. In a dynamic context, the ability to model economic disequilibrium is obviously important but it should be regarded as equally important in a (notionally static) spatial context. Many urban and regional markets are at best imperfectly competitive and have no stable economic equilibrium. This inherent economic instability entails a dynamic instability which, obviously, can be captured only through dynamic modelling. In such conditions, microeconomics assumes a fundamentally important role because forecasting and adaptation become necessary. Recursive programming models are peculiarly well suited for representing these phenomena and the different assumptions about agent behaviour associated with them.

Some related approaches are also worthy of our attention. Kuenne has recently produced a collection of papers around the themes of spatial oligopoly, rivalrous consonance and crippled optimization (Kuenne (1992)). He uses an equilibrium framework but his ideas would lend themselves to dynamic disequilibrium modelling. Marris's work (op. cit.) on the simulation of imperfect competition is also capable of being converted into a spatial disequilibrium form. The much older, and sadly neglected, work of Murphy (1965) on adaptive control process representations of investment behaviour, including some interesting ideas on subjective entropy, is worth further consideration as well.

Each of these approaches contains interesting ideas about agent behaviour. It is reasonable to assume that the more complex these ideas become, the more computationally taxing the operationalization of the theory will be. But this is not necessarily so. Indeed, one of the main justifications for employing a microeconomically sophisticated disequilibrium setting is that the need to separate the theory from the solution algorithm can be reduced dramatically. The theory can be formulated in algorithmic terms so that the simulated agents solve their problems for themselves, as it were. The economically arbitrary search process in the algorithm for finding an equilibrium can be replaced by economically purposive search behaviour of simulated agents.

7.11 CONCLUSIONS

The obvious stimulus to develop dynamic approaches in spatial analysis is the unsatisfactory nature of the treatment of dynamics in contemporary models. It has been argued that much mainstream work has been static or comparative static, so explicitly dynamic approaches are certainly needed. It has also been argued that the recent interest in nonlinear dynamics is certainly not the first serious attempt to tackle this problem. Moreover, it has been claimed that certain earlier attempts have some advantages over current methods because they attack the roots of the empirical problems associated with dynamics whereas many of the current methods skate over them. In particular, current methods pay insufficient attention to the empirical problems surrounding the existence and stability of economic equilibria and the nature of disequilibrium decision making by economic agents.

The brief examination of the development of the spatial sciences in this paper suggests that there is one key lesson we should learn. In the same way that it was ill-advised in the past to presume that economic equilibria exist and build models accordingly, it is equally ill-advised to presume that spatial systems behave in catastrophic or chaotic ways and try to build models that will deliver that behaviour. A more satisfactory approach theoretically and, perhaps, computationally, would be to formulate good descriptions of individual behaviour and good descriptions of agent interaction, then let the simulated agents sort out prices, allocations and spatial interactions amongst themselves, as it were. It would be surprising if behaviour resembling catastrophes and chaos did not emerge. If it did, we would have not just a description of catastrophes and chaos but a ready-made and robust explanation.

REFERENCES

Allen, P. M., & Sanglier, M., 1981, Urban Evolution, Self-organisation, and Decision-making, **Environment and Planning A**, 13, pp. 167-184.

Anas, A., 1984, Discrete Choice Theory and the General Equilibrium of Employment, Housing, and Travel Networks in a Lowry-type Model of the Urban Economy, **Environment and Planning A**, 16, pp. 1489-1502.

Anas, A., 1987, **Modelling in Urban and Regional Economics**, Chur: Harwood Academic.

Batty, M., P. Longley, & S. Fotheringham, 1989, Urban Growth and Form: Scaling, Fractal Geometry, and Diffusion-limited Aggregation, **Environment and Planning A**, 21, pp. 1447-1472.

Clarke, M., & A.G. Wilson, 1985, The Dynamics of Urban Spatial Structure - the Progress of a Research-Program, **Transactions, Institute of British Geographers**, 10(4), 427-451.

Cromley, R.G., 1982, The Von-Thünen Model and Environmental Uncertainty, **Annals of the Association of American Geographers**, 72, pp. 404-410.

Day, R.H., & T. Groves, 1975, **Adaptive Economic Models**, Academic Press, New York.

Day, R.H., & E.H. Tinney, 1969, A Dynamic von Thünen Model, **Geographical Analysis**, 1, pp. 137-151.

Dendrinos, D. & H. Mullally, 1985, **Urban Evolution: Studies in the Mathematical Ecology of Cities**, Oxford University Press, Oxford.

Forrester, J.W., 1969, **Urban Dynamics**, MIT Press, Cambridge, Mass.

Harris, B., 1985, Synthetic Geography: The Nature of our Understanding of Cities, **Environment and Planning A**, 17, pp. 443-464.

Harris, B. & A.G. Wilson, 1978, Equilibrium Values and Dynamics of Attractiveness Terms in Production-constrained Spatial-interaction Models, **Environment and Planning A**, 10, pp. 371-388.

King, A.T., 1977, Computing General Equilibrium Prices for Spatial Economies, **The Review of Economics and Statistics**, 59, pp. 340-350.

Koopmans, T.C. & M.J. Beckmann, 1957, Assignment Problems and the Location of Economic Activities, **Econometrica**, 1, pp. 53-76.

Kuenne, R.E., 1992, **The Economics of Oligopolistic Competition**, Blackwells, Oxford.

Marris, R., 1991, **Reconstructing Keynesian Economics with Imperfect Competition: a desktop simulation**, Edward Elgar, Aldershot.

Mills, E.S., 1974, Mathematical Models for Urban Planning, (A. Brown et al., eds.), **Urban and Social Economics in Market and Planned Economies**, Praeger, New York.

Murphy, R.E. Jr., 1965, **Adaptive Processes in Economic Systems**, Academic Press, New York.

Nijkamp, P., & A. Reggiani, 1988, Analysis of Dynamic Spatial Interaction Models by Means of Optimal Control, **Geographical Analysis**, 20, pp. 18-29.

Reggiani, A., 1990, **Spatial Interaction Models: New Directions**, Ph.D., Vrije Universiteit, Amsterdam.

Takayama, T. & G. Judge, 1971, **Spatial and Temporal Price and Allocation Models**, North Holland, Amsterdam.

Williams, H.C.W.L., 1977, On the Formation of Travel Demand Models and Economic Evaluation Measures of User Benefit, **Environment and Planning A**, 9, pp. 285-344.

Wilson, A. G. & Birkin, M., 1987, Dynamic-models of Agricultural Location in a Spatial Interaction Framework, **Geographical Analysis**, 19, pp. 31-56.

Table 1. Notation for the comparison of the modified Day-Tinney and Wilson-Birkin models of agricultural land use through time.

DAY AND TINNEY	WILSON AND BIRKIN
<td colspan="2">**COMMON TERMS** **Time-independent:** $c_{ij}^k \equiv$ cost per unit of transporting k from i to j; $v_i^k \equiv$ cost per unit of producing k in i. $a_i^k \equiv$ land used per unit of output of k at i **Time-dependent:** $p_j^k(t) \equiv$ price of product k at market j in period t; $\hat{p}_j^k(t) \equiv$ forecast of price of k in j at t; $X_j^k(t) \equiv$ demand for k in i at t; $Y_{ij}^k(t) \equiv$ flow of k from i to j in t; $Z_i^k(t) \equiv$ output of k in i at t; $E_i^k(t) \equiv \sum_j (\hat{p}_j^k(t) - c_{ij}^k - v_i^k) Y_{ij}^k(t) \equiv$ expected net revenue from the production of i at k. [in the notation used to describe the Harris & Wilson approach, $E_i^k(t) \equiv D_i^k(t) - C_i^k(t)$]</td>	
SPECIFIC TERMS $r_i \equiv$ overheads in i λ_j^k, μ_j^{km} are price function parameters $L_i \equiv$ land available at i $K_i(t) \equiv$ capital available at i in period t	**SPECIFIC TERMS** $r_i(t) \equiv$ land rent in i at t θ_i^k, η_i^{km} are demand function parameters $\varepsilon_1^m, \varepsilon_2^k, \delta_j^{mk}$ are adjustment parameters $\beta^k \equiv$ distance decay parameter for k $A_j^k(t) \equiv \left(\sum_i p_j^k(t) X_j^k(t) \exp(-\beta^k c_{ij}^k) \right)^{-1}$ (spatial interaction balancing term)

Table 2. Comparison of the modified Day-Tinney and Wilson-Birkin models of agricultural land use through time.

DAY AND TINNEY	WILSON AND BIRKIN
ALLOCATIONS AND SPATIAL INTERACTIONS For $t = 1,...$ find $Y_{ij}^k(t)$, $Z_i^k(t)$ $\forall ijk$ to maximize $$\sum_{ijk} E_j^k(t) Y_{ij}^k(t)$$ s.t. $$\sum_j Y_{ij}^k(t) - Z_i^k(t) = 0 \quad \forall ik$$ $$\sum_k a_i^k Z_i^k(t) \leq L_i \quad \forall i$$ $$\sum_k \left(c_{ij}^k + v_i^k\right) Z_i^k(t) \leq K_i(t) \quad \forall i$$ $Y_{ij}^k(t), Z_i^k(t) \geq 0 \quad \forall ijk$	**ALLOCATIONS AND SPATIAL INTERACTIONS** For $t = 1,...$ find $Y_{ij}^k(t)$ $\forall ijk$ to maximize $$\sum_{ijk} \left(\frac{1}{\beta^k} \ln Y_{ij}^k(t) + \frac{1}{\beta^k} \ln p_j^k(t) X_j^k(t) - c_{ij}^k\right) Y_{ij}^k(t)$$ s.t. $$\sum_j Y_{ij}^k(t) - Z_i^k(t) = 0 \quad \forall ik, \text{ where } \forall ik,$$ $$a_i^k Z_i^k(t) = a_i^k Z_i^k(t-1) + \sum_{m \neq k} \delta_i^{mk}(t-1) \varepsilon_1^m a_i^m Z_i^m(t-1)$$ $$\delta_i^{mk}(t) = \begin{cases} 1 & \text{if } E_i^k(t) > E_i^n(t) \; \forall n \neq k \\ -1 & \text{if } E_i^m(t) > E_i^n(t) \; \forall n \neq m \\ 0 & \text{otherwise} \end{cases} \forall ik$$
EXPECTED PRICES: Naive forecasts $\hat{p}_j^k(t) = p_j^k(t-1) \quad \forall jk$	**EXPECTED PRICES: Perfect foresight** $\hat{p}_j^k(t) = p_j^k(t) \quad \forall jk$
DEMAND AND PRICES $p_j^k(t) = \max\left\{0, \left(\lambda_j^k + \sum_m \mu_j^{km} X_j^m(t)\right)\right\} \quad \forall jk$ $X_i^k(t) = \sum_j Y_{ij}^k(t) \quad \forall jk$	**DEMAND AND PRICES** $X_j^k(t) = \max\left\{0, \left(\theta_j^k - \sum_m \eta_j^{km} p_j^m(t)\right)\right\} \quad \forall jk$ $p_j^k(t+1) = p_j^k(t) + \varepsilon_2^k\left(E_j^k(t)\right) / X_j^k(t) \quad \forall jk$
RESIDUAL INCOME: used as capital $K_i(t+1) = \sum_{jk} p_j^k(t) Y_{ij}^k(t) - r_i \quad \forall i$	**RESIDUAL INCOME: paid as rent** $r_i(t) = \sum_{jk} E_i^k(t) Y_{ij}^k(t) \quad \forall ik$

CHAPTER 8
SPATIAL EVOLUTION IN COMPLEX SYSTEMS
Peter M. Allen

8.1 INTRODUCTION

Understanding the emergence and evolution of structure and pattern in natural and human systems is one of the most important, and complex questions that can be addressed. The transition from simply describing and mapping spatial patterns, to understanding how they emerge and how they can change over time has proved to be difficult. In general we can say that the latter questions did not even begin to receive a mathematical framework within which answers might be sought, until the study of dissipative structures (Nicolis and Prigogine, 1977, Prigogine, Herman and Allen, 1977) and synergetic systems (Haken, 1977) began some 15 years ago.

The reasons are deeply rooted in the history of science, where the equilibrium, and mechanistic paradigms reigned supreme for many years. Rational analysis and reductionist science were responsible for the great technological and scientific advances of the 19th and early 20th centuries, leading to the industrial revolution and to enormous improvements in living conditions and this success made it difficult for new ideas concerning complex systems to be accepted.

Over recent years, however, these new ideas concerning structure and organization in non-equilibrium systems have become better known. Advances in our understanding of the behaviour of nonlinear systems has led to a new view of evolutionary processes, which essentially modifies the traditional paradigm of science, and presents us with a new basis of understanding structure and organization and its change over time.

In this paper we shall present these new ideas, and show the importance and utility in practical problems of urban and regional structure, and in the search for sustainable development.

8.2 THE MECHANICAL PARADIGM

The basis of scientific understanding has traditionally been the mechanical model (Prigogine & Stengers, 1987; Allen, 1988). In this view, the behaviour of a system could be understood, and anticipated, by classifying and identifying its **components** and the **causal links**, or mechanisms, that act between them. The object under study is represented by a "machine" that can be run forward to provide predictions, and whose component variables reflect the taxonomy of the system. In many cases such systems were also supposed to have run themselves to equilibrium, so that the correspondence between the real object and that model was made through equilibrium relations of balance between the variables. In geography, for example, urban form and hierarchy was supposed to express some maximised utility for the actors, where consumers had minimised distance of travel for goods and services and producers had maximised profits. Such ideas gave rise in reality to a purely descriptive approach to such problems, but as computers became more powerful, the need for such strong and unrealistic assumptions gradually disappeared. System Dynamics and simulation were thought to offer a path to the prediction of system behaviour, and because of this to offer a basis for rational policy and decision making in complex systems.

In recent years, therefore, much attention has rightly been given to the interesting behaviours of nonlinear dynamical systems. Such systems are characterized by dynamical equations of the type:

$$\frac{dx}{dt} = G(x, y, z....)$$

$$\frac{dy}{dt} = H(x, y, z....)$$

$$\frac{dz}{dt} = J(x, y, z....)$$

where G, H, and J are functions which have nonlinear terms in them, leading to changes in x, y and z which are not simply proportional to their size. Also, these functions are made up of terms which involve variables x, y and z and also parameters expressing the functional dependence on these. These parameters reflect two fundamentally different factors in the working of the system:

- the values of **external** factors, which are not modelled as variables in the system. These reflect the "environment" of the system, and of course may be dependent on spatial coordinates. Temperature, climate, soils, world prices, interest rates are possible examples of such factors.
- the values corresponding to the "performance" of the entities underlying x, y or z, due to their **internal** characteristics like morphology, level of knowledge or strategies.

These two entirely different aspects have not been separated out in much of the previous work concerning nonlinear systems, and so the whole issue of the evolution of the populations involved in the system has not been addressed clearly. Equations of the type shown above display a rich spectrum of possible behaviours in different regions of both parameter space and initial conditions. They range from a simple approach to a homogeneous steady state, characterized by a point attractor, through that of sustained oscillation of a cyclic attractor, to the well known chaotic behaviour characteristic of a strange attractor. These can either be homogeneous, or can involve spatial structure as well. This possibility of rich behaviours has proved to be of great significance for many fields of science.

However, in this paper, we wish to draw attention to the fact that there is another important extension that can be made to such mathematical models: that of modelling systems in which **adaptive** and **structural** change can occur. This corresponds to the problem of evolutionary systems, where new variables and new mechanisms of interaction appear spontaneously from within the system itself, leading to a changing taxonomy. The picture that we have is that of a hierarchy of structures, in which the parameters of "performance" perceived at the macroscopic level are actually achieved by spatial reorganizations at the microlevel of interacting individuals, groups and populations.

Many problems are studied using kinetic equations which operate at a macroscopic level, and which describe the actions of different mechanisms operating between microscopic components. For example, hydrodynamics describes the average effects of molecular interactions in a fluid, chemical kinetics the average effects of molecular chemical interactions, population dynamics the average effects of birth and death processes and "system dynamics" the average change in some general system as a result of the processes of increase and decrease operating between its elements.

Typically, we have for example the logistic equation describing the change in population x over time as a result of birth and death terms in a limited environment.

$$\frac{dx}{dt} = bx(1 - \frac{x}{N}) - mx$$

It is really only written correctly in terms of density of population. It can be derived from the underlying reality of discrete events such as birth and death by observing the average number of births and deaths per unit time over some volume or area and using this to "predict" the dynamics of the population. This equation is typical of many geographical models which use population dynamics, and therefore the remarks below apply equally to these.

As has been shown elsewhere (Allen, 1990), however, in order to derive this equation two assumptions are required:
- all individuals are identical and of average type
- only the most probable events occur.

The errors introduced by the second assumption can be addressed by using a deeper, probabilistic dynamics, called the "Master Equation" which essentially assumes that all individuals are identical and equal to an average type, but that events of different probabilities can and do occur. So, sequences of events which correspond to successive runs of good or bad "luck" are included, with their relevant probabilities. In the case of the logistic equation, this richer evolution does not change the result very much, but if the equation is more nonlinear, and more than one stationary state is possible, then the probabilistic evolution shows that the concept of a deterministic trajectory coming from the kinetic equation view, is misleading.

The fuller description shows us that a real system can "tunnel" across the separatrix between basins of attraction, and that it can therefore flip occasionally between seemingly stable states. Similarly, in a region of parameter space corresponding to chaotic motion, clearly, a real system will not behave according to its deterministic kinetic equation, since it will be continuously perturbed from its trajectory by microscopic fluctuations. However, the probability distribution will still wrap itself around the surface of the strange attractor, giving the same average outcome as the kinetic approach. These issues have been discussed and presented elsewhere (Allen 1990).

What is of interest here is the problem presented by the other hypothesis that all individuals are **identical and equal to the average type.** Obviously, the first and most important fact about two individuals is that they cannot be at the same place at the same time, and so spatial structure is one of the underlying issues that has been treated inadequately by population dynamics of the usual kind. In the next section we show the result of correcting nonlinear equations in order to take the effects of microscopic diversity into account.

8.3 EVOLUTIONARY DRIVE: MODELS OF ADAPTATION

From the description above, it is clear that although the Master Equation addresses one of the inadequacies of the kinetic approach, that is it does not assume that only the most probable events occur, it still assumes that the population concerned is made up of identical individuals, of average type. More precisely, it assumes that individuals are sufficiently similar to each other to be characterized by the same parameter values relating to the mechanisms that "increase" and "decrease" that population. At first, this seems to be simply a condition on the choice of variables for the model, which is based on the taxonomy that has been established, where the different populations that are in interaction have been identified as being made up of individuals who are sufficiently similar.

Now let us examine the problem of allowing for the fact that in any population or category, individuals will **not** be identical. In other words, our models must take into account non-average behaviour, and also the possibility that at the microscopic level of individuals, error making processes of reproduction or of behavioral imitation provide a mechanism for the maintenance, decrease or possibly the increase, of diversity within the system, leading potentially to a change in the observed taxonomy.

In order to demonstrate this idea as clearly as possible, let us consider the logistic equation above. Instead of the simple kinetic equation form above, let us assume that there is some dimension of the character or behaviour space of x, i, in which the population is distributed. This might be colour, size, spatial position, morphology, or behavioral particularity. The behaviour of any single particular sub-population will be:

$$\frac{dx(i)}{dt} = b(i)x(i)[1 - \sum_{i'} \beta_{i'i} x(i')/N(i)] - m(i)x(i)$$

where we can admit that the rate of reproduction b(i) and of mortality m(i) may be some function of the characteristic i. The value of N(i) represents the size of the "niche" open to the behaviour/characteristic i, and the $\beta_{i'i}$ take into account the competition between type i' and i for the same niche.

However, more realistically, we must suppose that there can be a "diffusion" of x between the different types of behaviour. This may either occur through attempts at imitation of seemingly successful strategies, or through "errors" in the handing on of "character" to successors. We assume for example that reproduction is slightly imperfect and produces a small fraction of off-spring which are in the neighbouring boxes, i+1 and i-1. Let us suppose that f is the "fidelity" of reproduction, and that the mechanisms which "pass on" the necessary behaviour over time are quite successful, possibly 95% for example, but that in that case 5% of reproduction goes into the neighbouring types. This equation is somewhat similar to the master equation, but here it is the "error making" that links the different slices of the distribution of individual types. If we assume that all the different population types are subject to the same limiting factor, then we find that there are a set of equations of the form:

$$\frac{dx(i)}{dt} = \{fb(i)x(i) + .5(1-f)b(i+1)x(i+1) + .5(1-f)b(i-1)x(i-1)\}\{1 - \frac{Totpop}{N(i)}\} - m(i)$$

where totpop is the total population of all x: totpop $=\Sigma_i$ x(i), and N(i) is the limiting factor for behaviour i.

In general then, if the i space corresponds to different types of x, their performances with respect to "birth" and "death" may differ, and give rise to differential performance.

If we choose some functional form for b(i) and m(i) so that there is an evolutionary landscape, then the distribution adapts over time, climbing the slope of evolutionary advantage (Allen and McGlade, 1988, 1989).

In a competition between a population of "error makers" and a population of perfect replicators, the former out-compete the latter in the long run, and evolution selects for the capacity to adapt and learn rather than simply for good, but fixed behaviour. Such a model is therefore a simple example of a new class of models - ones that contain an endogenous capacity to evolve the nature of the populations that participate in the system.

In as much as "spatial location" might be considered as a characteristic of an individual, the work on "dissipative structures" of Prigogine and coworkers and the "synergistic structures" of Haken and his co-workers already represents a first step in this direction. It was shown that nonlinear interactions could spontaneously structure systems and that this provided the foundations for an understanding of morphogenesis in many fields. In both these groups a wide range of applications of these ideas were undertaken, for example in the realms of ecological and human systems modelling, but generally concentrating on the spatial dimension of "difference" between individuals in a population. In this way, Deneubourg showed how insect societies could use these principles to build spatial structures such as nests, and lay trails to food sources. In a series of papers (Allen, 1978, 1982, 1983, 1984; Allen & Sanglier, 1979, 1981; Allen, Engelen and Sanglier, 1983; Allen, Sanglier, Engelen and Boon, 1985; Sanglier and Allen, 1989), showed how settlement patterns, spatial geography, urban hierarchies and urban structure could all be described by the same kind of self-organization process in the spatial dimension. In that case, the "urban multipliers" of accelerated economic activity and job opportunity amplified differences and led to the geographical organization of the system. In general the process of morphogenesis can be viewed as the co-evolution of spatial gradients and the patterns of concentration of the constituent populations.

Eigen and Schuster (Eigen and Schuster, 1979) introduced the idea of selection among different types of polymer and demonstrated how life could begin spontaneously from this interplay of variation and nonlinear dynamics. Their work is therefore also an important aspect of the general principle that we wish to introduce here. That is that if we simulate evolution, it selects for populations that do not translate behaviours perfectly through time, but which explore neighbouring behaviours constantly through the maintenance of microscopic diversity.

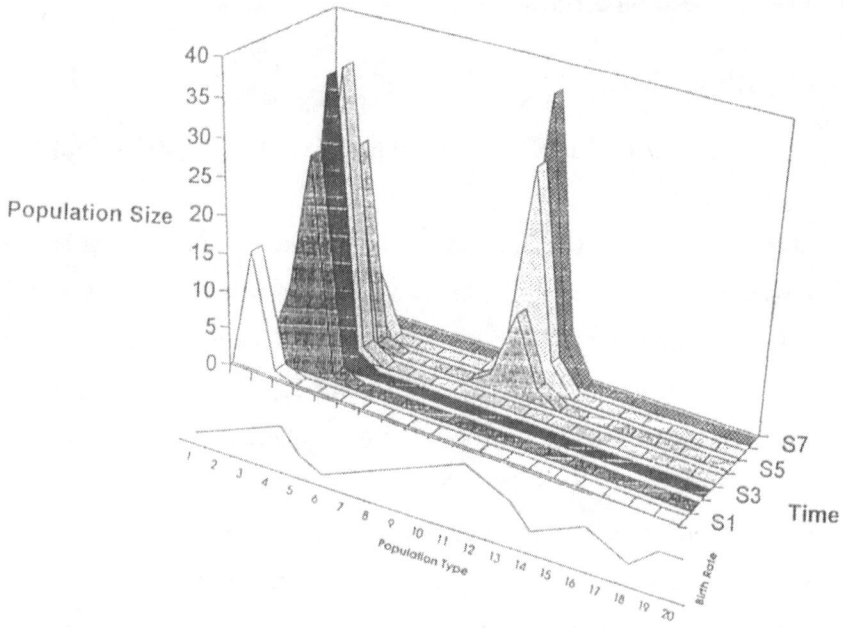

Figure 1   Simple evolutionary landscape, which the microscopic "error making" of the x, drives the "average" x up the slope.

From the simple example indicated above, several experiments have been made which show that complex systems are made up of populations with the capacity to adapt their identities to the changing "landscapes" of opportunity. In the case shown above the slope of the landscape of advantage was **fixed** and was neither affected by the movement of the population up the slope, nor by the adaptive responses of the other populations with which X is in interaction. But, this can also be explored by such models so that the shape of the adaptive landscape depends on the behaviours of the other populations present. For example, mutually evolving predator/prey systems were studied already in 1976 (Allen), but now instead of leaving "variability" as an external perturbation of the equations, we now can internalize it. The dimension of "adaptive landscape" of the prey and the predators will reflect the different types of characteristic that affects the parameters of interaction. For the prey these will be factors affecting the ability to find and use food, to escape detection or capture by the predator, and to reduce death due to other causes.

These may involve for example camouflage, speed, silence, and reproductive rate. For the predator, the space will involve those factors which affect its ability to detect and capture prey, and to avoid death at the hands of other predators.

8.4 EVOLUTION AND EMERGENT ECOLOGICAL ORGANIZATION

The incorporation of non-average behaviour in the dynamics of a system has led to a simple vision of populations adapting to either a fixed landscape or to another population. But, here we can introduce another richer idea which is that populations adapt to each other. That is, the environment of individuals within a population is actually made up of the **external** environment, the environment of **other populations**, and also the other members of the **same** population (Allen 1990, Allen and Lesser, 1991). The performance of an individual is judged in a complex manner therefore, in part as one of a growing or shrinking species in the external environment and system, and also with respect to the other members of the same population. This question was also examined in an earlier paper, but again the generality of the reasoning was not clearly demonstrated. In a further experiment we can write a single equation for a population in a space of possible character or behaviour, such that the different types would access different resources to some degree. In other words, depending on a parameter of resource type gradient, we may say that populations of neighbouring type compete more for the same resources than if they were further apart in the character space. The equation for this is:

$$\frac{dx_{(i,j)}}{dt} = \left((fbx_{(i,j)} + wx^2_{(i,j)}) + \frac{b(1-f)(x_{(i+1,j)} + wx^2_{(i,j)})}{4}\right.$$

$$+ \frac{b(1-f)(x_{(i-1,j)} + wx^2_{(i-1,j)})}{4} + \frac{b(1-f)(x_{(i,j-1)} + wx^2_{(i,j+1)})}{4}$$

$$\left. + \frac{b(1-f)(x_{(i,j-1)} + wx^2_{(i,j-1)})}{4}\right)(1 - \frac{\sum_{(i',j')} x_{(i',j')} e^{-pd(i,j:\hat{i}\hat{j})}}{N\sum_{(i',j')} e^{-pd(i,j:\hat{i}\hat{j})}}) - mx_{(i,j)}$$

Now, by running equation (8.2) we can show that over time a single population generates a simple ecology of populations, as shown in Figure 2, where each one is "held" in place by the presence of the other populations that emerge over time. The taxonomy of the system is changing as it runs, and there are three types of qualitative evolution occurring: i) the taxonomy is stable, and the "ancestral" branches can suppress "error making" explorations around them. ii) the "ancestral" lines can branch due to internal competition when error makers are amplified, and iii) entirely new populations can emerge in spontaneously in the unfilled regions of character space.

This simulation shows us that the kinetic equations used for many studies of nonlinear dynamic systems are essentially incomplete. They merely capture the **functioning** of a system based on the taxonomy evident at the initial time. If all the underlying microscopic components are indeed identical, and remain so, then the dynamics will give a correct prediction over time, except at bifurcations when the underlying master equation needs to be solved in order to describe correctly the precise course of events. However, of greater concern here is the fact that owing to the presence of **non-average** individuals the taxonomy of the system can change in the manner shown above. The presence of "error making" in reproduction or in the transmission of behaviour patterns or strategies mean that there is a **creative** presence in the system which drives adaptation and structural evolution in an inherently unpredictable direction, in that the properties of the emergent behaviour cannot be fully anticipated before they actually experience the system.

The study of "chaotic dynamics" although of great interest is only in fact dealing with a mechanical set of equations which do not take into account the fact that, in reality, the variables may change qualitatively over time. Dissipative structures, and synergetic systems however, have recognized the importance of "self-organization", and have concentrated on spatial structure and organization. This is indeed of great significance, but in addition to this we can now recognize the more general principle that self-organization can occur in the many dimensional space of the microscopic diversity. We have a kind of "reaction diffusion" system operating in the parameter space of the populations reflecting their identities, and the taxonomy underlying the system.

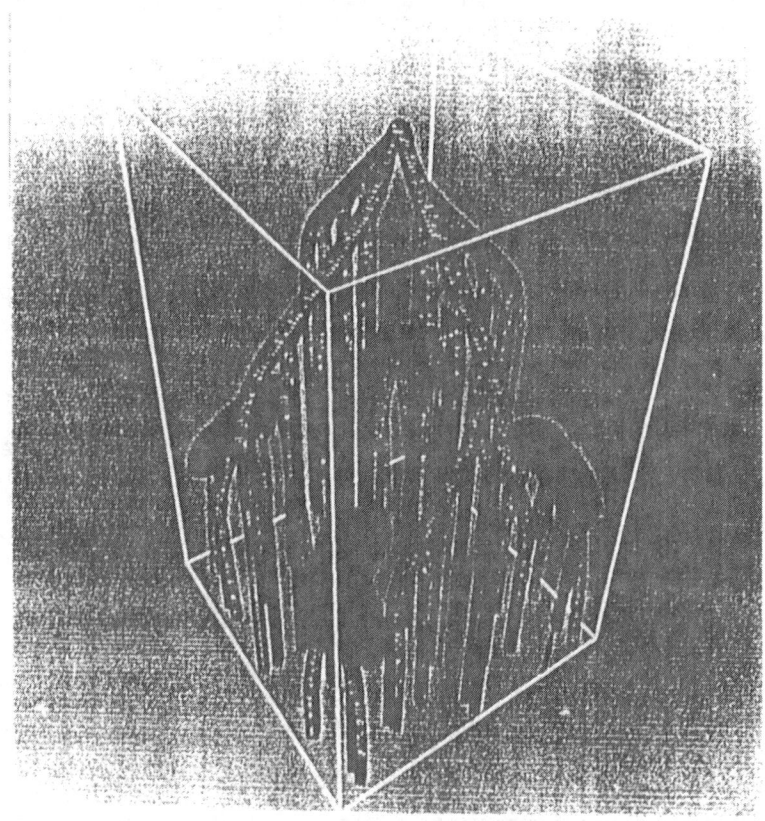

Figure 2   An emergent ecology of populations resulting from the interplay of error making and material constraints.

The mathematical models briefly introduced above suggest that the organizational principle of an ecosystem, and therefore of sustainable and resilient behaviour, is that of a level of microscopic "error making", corresponding to imperfect and subjective information, operating below an apparently functional macrostructure. Our models show us that evolution selects for this adaptive capacity hidden below the macroscopic level, and therefore that successful systems in the real world will have a wide diversity of behaviours present, much of which will be sub-optimal by any particular criterion. Scientific **explanation** of individual behaviours or characteristics may be impossible in such systems therefore, since the adaptive capacity that evolution leads to, requires that a spectrum of behaviours be there, and that "error making" should be possible, but does not specify necessarily that **particular** individual behaviours should be present. This leads to a curious inversion of the idea of the "Selfish Gene", whereby evolution may perhaps only explain the diversity of the ecosystem, but not the particular individual expressions of this.

The view advanced here shows us a unified conceptual framework for understanding and modelling complex systems. For systems with a fixed taxonomy where the discrete nature of the underlying microscopic events can be shown to be unimportant, nonlinear kinetic equations can be used, and may lead to different behaviours such as point, cyclic or chaotic attractors. The reality of discrete events however, may well require the use of a "Master Equation" approach. However, for open systems in general, we cannot assume that the taxonomy is fixed, and we must attempt to model the evolution of the different populations within the possibility space that we can imagine to be relevant. Of course, we cannot necessarily imagine correctly the possibility space of the component populations and because of this we must admit that there is a fundamental limitation on the possibility of scientific prediction.

Clearly, however, the evolutionary principle demonstrated above applies to the dimensions which affect the "performance" of individuals in the system, and thus which can lead to changes in the values of the parameters characterizing the endogenous properties of the population. However, geographic space is also an important factor that must also be considered, since it can also be viewed as a part of the endogenous character of an individual. Geographical space therefore plays a unique role in the system in that it may not only be important for the exogenous parameters such as climate, temperature etc. but also "location" can be seen as an intrinsic property of individuals, so that spatial structure and organization of a system can also result from the interplay of different spatial strategies.

8.5 SPATIAL EVOLUTION OF COMPLEX SYSTEMS

In the very first applications of these ideas to human systems (Allen & Sanglier, 1979, 1981), a nonlinear dynamical system of equations expressing the supply and demand of different products was made to evolve by the random occurrence of entrepreneurs at different points and times in the system. The supply side was characterized by a non-convex production function for different economic activities, and consumer demand was assumed to reflect relative prices. The creative interplay of random exploration and rational selection was simulated by the random parachuting of entrepreneurs onto the plain of potential demand which resulted in the gradual emergence of a stable market structure and pattern of settlement as economic functions either prospered or declined at their locations. The equations were of the form:

$$\frac{dx(i)}{dt} = bx(i) - mx(i) + \text{Mob}(i) \sum_j x(j) \cdot \frac{A_{ij}}{\sum_{i'} A_{i'j}} - \text{Mob}(i)x(i) \frac{\sum_j A_{ji}}{\sum_j A_{ji}}$$

The structure emerges as a result of the interplay of positive and negative feedbacks:
- positive due to the urban multiplier, economies of scale and externalities.
- negative through spatial competition both for producers and for residential space.

The system can get itself into what has been called "positive feedback traps"(Allen 1991), where the system can get itself locked into a somewhat unsatisfactory market structure, as a result of a particular history. But what these kind of simulations show us is that a very large number of possible stable structures could potentially result from the experiment, involving different numbers of centres in different locations, and necessarily not offering the same level of cost, utility, or efficiency. In some ways they show rather clearly that a "free market" system does not necessarily run to an "optimal" solution, but just to one of many possible solutions. The "invisible hand" of Adam Smith therefore, although ensuring that incompatible entities do not cohabit the space, cannot be left alone to guide the system, if we wish to attempt to achieve any goals, and avoid certain problems. Choices exist, but we can only successfully make the choices that we wish if we can understand the qualitative evolution of the system using these kind of model.

Even in this very preliminary form the models already showed many important principles: many final states are possible, precise prediction in the early stages is impossible, approximate rules appear (centre separations etc.) but always with considerable deviation and local individuality present. The results are affected by the particularities of the transportation system, as well as by information flows affecting the mental maps of consumers (Gould & White, 1978). Also, the evolution of structure as a result of changing technology, transportation, resource availability etc. can be explored, as the changing patterns of demand and supply affect each other in a complex dynamic spatial process.

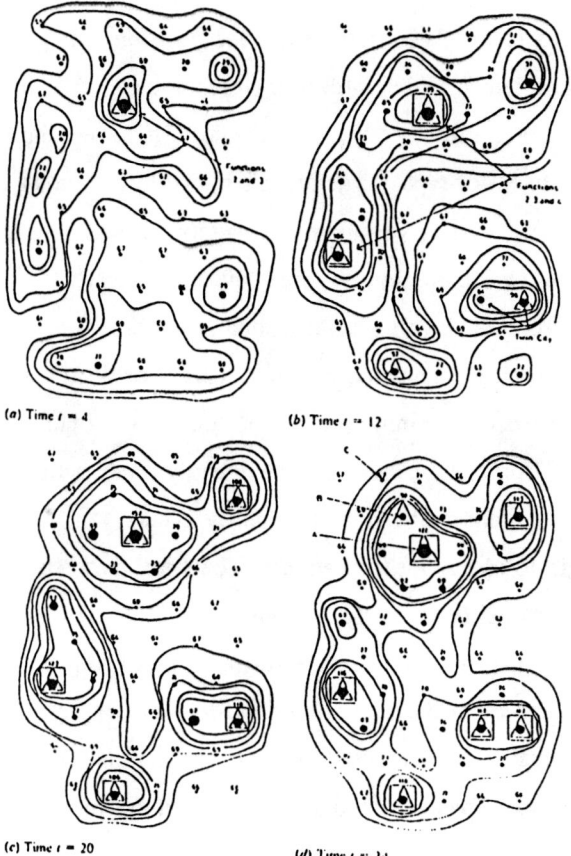

Figure 3   As economic functions are launched into the system at random locations they modify the spatial pattern of demand and supply in a complex coevolution.

As discussed earlier, these ideas were developed in a series of papers showing how the ideas of complex system evolution could be applied to urban, regional and national spatial systems (Allen, Allen et al. 1979-1989).

What is important is that these models contain the two essential ingredients that we have identified above for creative self-organization:

- there is a dialogue between random explorations and diffusion, and the rational forces of selection.
- there is a clear representation two levels of description, the individuals making up each category of actor, and the macrostructure of urban and regional form that they both generate and inhabit.

The fundamental basis for these models are the decisions of the different types of individual actors considered, which reflect their values and functional requirements. These may be very represented by very simple rules. However, the models show that even simple average rules for types of actor, when distributed among average and non-average individuals, gives rise to very complex patterns of structure and flow, and to a structural emergence and evolution at the collective level. In such systems the microscopic and macroscopic levels are not related in a simple fashion. It is not true that the large structure is simply the small writ large. This is because macroscopic structure emerges, and this affects the circumstances of the microscopic parts, as they find themselves playing a "role" in a larger, collective entity. Each actor is co-evolving with the structures resulting from the behaviour of all the others.

We can now begin to see a closer relationship between these spatial models and the general evolutionary model of the early sections.

Let us consider how the two distinctive features corresponding to "error making" and "rational selection" appear in the spatial models. In representing the decision making behaviour of individuals we have used the idea of "attractivities", which assign a probability to each of the choices open to an individual of a given type, located at a given point. The form that we use for an actor of type k is:

$$\text{Probability of choice}(i) = \frac{A_i^k}{\sum_j A_j^k}$$

where A_i^k is the attractivity of choice i as viewed by an actor of type k. The choice can concern the choice of sources of supply either for a private consumer, or for a producer looking for supplies of components. In that case it would probably reflect the cost to the consumer of the different choices, involving the prices practised by the producer and also the costs of transportation. However, it will also reflect the information of the actor concerning possible sources of supply, and this will show a dependence on purely personal information networks as well as on the "objective" truth. The attractivity in question may also concern places of residence and will reflect the location of the workplace of the actor concerned, as well as their personal knowledge and contacts. In other words, our spatial models take into account the internal diversity of populations of actors, which is exactly the issue underlying "evolutionary drive" that is described in the sections above. Because of this individual particularity of the spatial decision-making it means that a population of actors will fraction among the possible choices, so that even the most improbable choices will have some small number of actors trying them out. This in turn means that, if conditions change, and the "pay-offs" of a choice should increase, then our model will register an "adaptive" response of the actors, as more adopt that choice.

From this, we can see that "error making" is represented in these spatial models by the use of relative attractivities which take into account the non-average as well as the average behaviour of each type of actor.

The "selection" process on the other hand results simply from the success or failure of enterprises in the competitive and cooperative dynamical game that is running. Models were developed showing the evolution of cities and regions as a result of these evolutionary processes. For example, the generic interaction scheme of figure (4) was shown to successfully generate the emergence and evolution of the spatial structure of the city of Brussels.

In this model, the decisions of individuals actors of different types is simulated as each of them changes the conditions which affect the others. As the city grows, so the demand for space changes the price of available land, and the different abilities of actors to pay affects their responses. This difference reflects the different spatial requirements of economic activities, where for example, manufacturing industry may require twenty times the surface area as an insurance office, for the same value added. Also, different actors will have different requirements that they wish to fulfil with their location, such as either being very central for communication and contacts, or needing transportation infrastructure such as a port, rail and road access for the import raw materials and the export of finished products. Residents too will have different locational needs depending on where they work, and on there age, family composition and socio-economic group. Through the spatial interaction of all these decisions, running over time, the city structures and generates business and administrative districts, industrial satellites, retailing centres, and areas of high and low quality residential housing. All the spatial choices of the different actors are made according to the networks of private and public transportation, which will themselves be changed according to the intensity of demand and congestion that occurs. The models therefore were able to show the complex relationship between urban form and structure on the one hand and the transportation system and energy costs on the other.

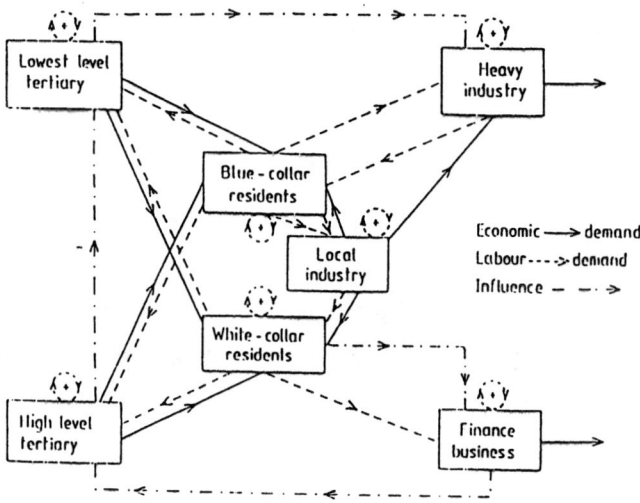

Figure 4   The intra-urban interaction scheme which leads to the development of urban structure.

The impact of a new metro line for example showed how the location of blue and white collar residents and commercial property were affected over the whole public transportation system. Also, the model was used to explore the possible impact of information systems and telecommunications, and it was shown to lead potentially to the dispersal of much of the office employment in the centre of the city, and therefore to a radical change in urban structure and in the pattern of use of urban transportation.

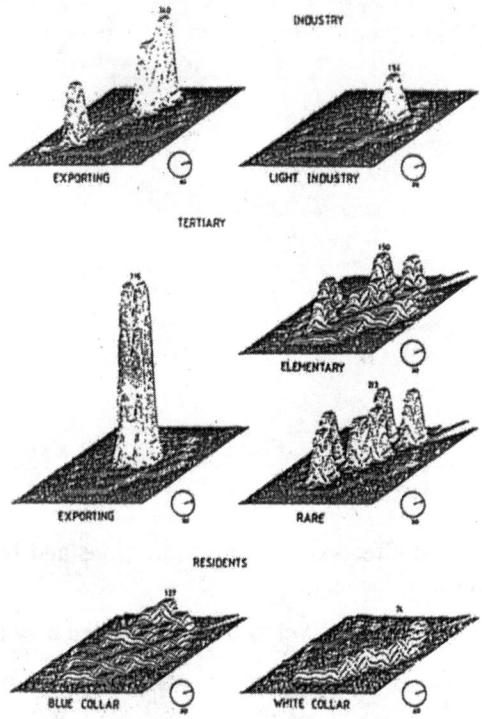

Figure 5   The city of Brussels that emerges from our interaction system after 20 units of time starting from a central seed.

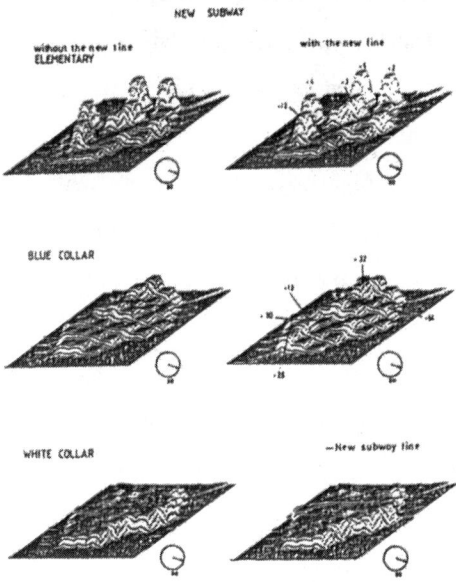

Figure 6   The effects of a new Metro line on some of the variables of the system.

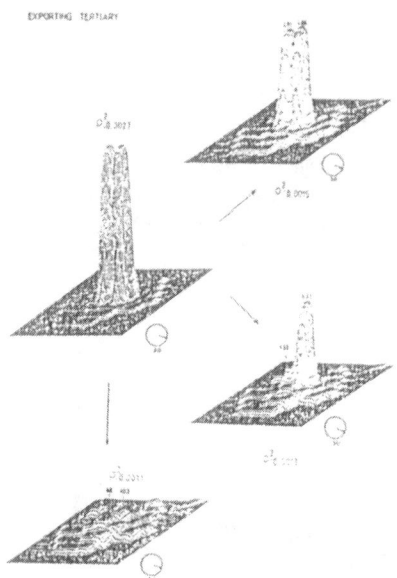

Figure 7   The possible effects of telecommunications and informatics on the office employment in the centre of the city.

Similarly, regional models were developed which showed how the urban centres within the system interacted. This mutual interaction was both influenced by, and in its turn influenced the flows of investment and of migrants. From this an evolutionary model of Belgium, for example, was developed and has been described in detail elsewhere. What is important, however, is that the overall effectiveness of a city is conditioned by the costs and benefits that result from its location within the national framework, but also from the structure of its component cities. So, cities that suffer high levels of congestion, have poor infrastructure, dissatisfied residents with poor educational and training levels, environmental problems and high taxes, for example, will not be characterized by the same parameters of "functioning" as they would be if they had better internal structures and facilities. Because of this, the particularities of the internal structure, and the success or failure of urban centres with respect to their inhabitants, penetrates upwards to affect their capacity to attract investment and migrants, and hence their long term growth.

What emerges, is a hierarchy of cities and towns, where the potentials and opportunities are affected by the global performance of the nation as a whole, by their relative location and performance parameters within the national system, and these latter in turn are affected by their internal structures. Conversely, the internal structures are affected by the relative success of the city within the nation, and the nation within the international system. The micro-levels are linked in a circular causality to the macrolevels.

The spatial models of urban and regional evolution therefore demonstrate the fact that what emerges are **ecologies** of populations, clustered into mutually consistent locations and activities, expressing a mixture of competition and symbiosis. The effectiveness of a region depends on the collective linkage of the towns and cities that exist within it, as well as on the performance of the cities and towns which make it up, and these in turn on the structures that emerge for different neighbourhoods. Individuals, either as private citizens or in their roles as decision makers concerning the location of economic activities, and the investments that ebb and flow to them, both respond to and influence the structures they inhabit. This nested hierarchy of structure is the result of evolution. It is not "optimal" in any way, because there are a multiplicity of subjectivities and intentions, fed by a web of imperfect information. However, this complex, lack of optimal organization is what allows adaptive responses to occur to changing conditions and opportunities. We have learnt the hard way that although "company towns", or highly specialized cities,can succeed remarkably for some time, in the end lack the resilience and adaptability for permanent survival. This can only come from diversity, and from the constant maintenance of the "means" of growing new businesses.

The spatial choices for each type of actor are represented by an "attractivity" term as described above, but this term takes into account not only the average behaviour, which is the most "rational" one for that kind of actor, but also the non-average behaviour. No choice gets a zero probability allocation, however seemingly stupid it is as a choice. The probability may be small, but depending on the heterogeneity, and the level of rationality in the population, there will always be some individuals who make a different choice for their own reasons, or lack of information. Because of this, the total pattern emerges as a result of the interaction of imperfect patterns of behaviour for each type actor, and what this really means is that there is an intrinsic element of unpredictability in the system, since the non-average deviations that will **actually** occur cannot be predicted on rational grounds. Positive feedback can therefore amplify these unpredictable deviations and lead to the creation of new structure. Creativity, and adaptive response are therefore powered by the degree of heterogeneity of the population, and their microscopic diversity.

8.6 DISCUSSION

This new approach has been continuously developed since 1978, in order to provide realistic tools for understanding regional and urban evolution, in which the patterns of structure and flows are the result of an on-going evolutionary process of self-organization. In the papers cited earlier, applications of these ideas have been made in the USA, Belgium, Holland and France, as well as in understanding intra-urban evolution in cities of Belgium and France.

The most complete and latest version is designed to help in the formulation of development policy for Sénégal. It allows the user to change the transport network, energy costs, to develop new industrial or agricultural infrastructure, and to explore the consequences over time as the different links between demography, economic activities and the environment affect each other. It has a very user-friendly interface, which has an explanation facility, and an interactive help facility, as well as menu driven policy inputs, and comparison facilities. The migration flows over the long term have been studied and successfully generated over periods of 30 years. The migration models generate not only the pattern of flows at any given time, but also the evolution of these patterns as a result of economic growth and decline, changing factors such as soil productivity and water availability, and the changing pattern of demographic potential.

Let us now consider more fully the relationship between the evolutionary processes described in the sections 2 and 3 and the models of spatial evolution in section 4. The first comment is that the figure of ecological emergence clearly resembles very closely that of the spread of urban centres into a previously empty zone. This is because, if we choose the two dimensions of "behaviour" to be the 2-D of a landscape then figure (2) can be interpreted as the colonisation of geographical space through a simple diffusion process, which encounters and generates loops of positive and negative feedback. In other words, spatial evolution is just a particular case of the more general co-evolution of populations.

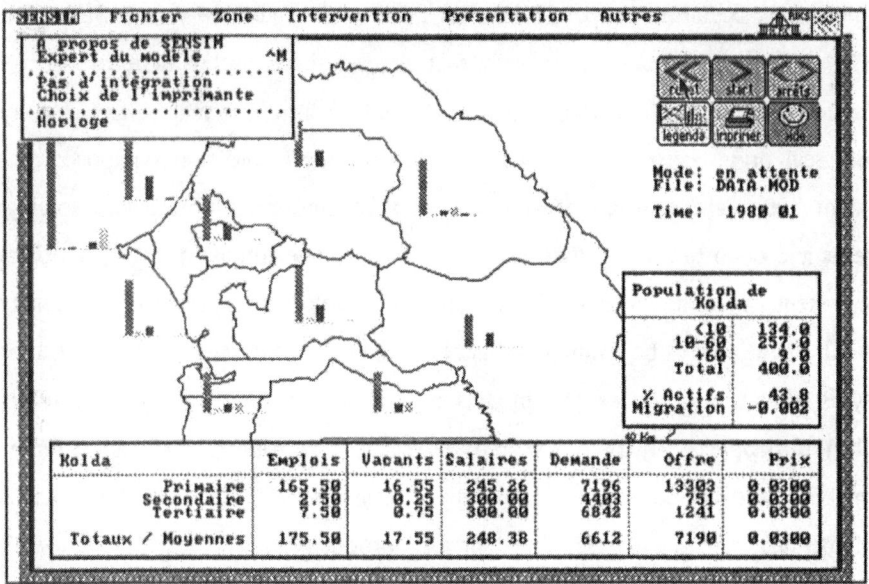

Figure 8. An interactive spatial model of Sénégal that can be used for policy exploration.

From this we can begin to make explicit a view about "sustainable" economic development, as being connected to the ecological and adaptive nature of a system, rather than its economic efficiency, or indeed any other simple optimality. Sustainability must be understood in terms of the ideas that have been developed here. It should not be interpreted as the search for the perfect equilibrium. The world will never stop changing, and what sustainability is really about is the capacity to respond, to adapt and to invent new activities. The power to do this lies not in extreme efficiency, nor can it be had by simply distributing cash. It lies in **creativity**. And in turn this is rooted in diversity, cultural richness, openness, and the will and ability to experiment and to take risks.

Instead of viewing the changes that occur in a complex system as necessarily reflecting progress up some pre-existing (if complex) "quality of life" landscape, we have shown that the landscape of possible advantage itself is produced by the actors in interaction, and that the detailed history of the exploration process itself affects the outcome. Paradoxically, **uncertainty** is therefore **inevitable**, and we must face up to this. Long term success is not just about improving performance with respect to a fixed external environment of resources and technology, but also is affected by the "internal game" of a complex society. The "pay-off" of any action for an individual cannot be stated in absolute terms, because it depends on what other individuals are doing. Strategies are **interdependent**, and spatial structures that emerge in complex systems are nested hierarchies with the feedback loops operating both from micro to macro, and vice versa.

Ecological organization is what results from evolution, and it is composed of self-consistent "sets" and patterns of activities or strategies, both posing and solving the problems and opportunities of their mutual existence. The future, then, is not contained in the present, since the landscape is fashioned by the explorations of its inhabitants, and adaptability will always be required. This does not necessarily mean that total **individual** liberty is always best, since our models also show that adaptability is a **group or population property**. It is the shared experiences of others that can offer much information, and of course, it is the division of labour, and the development of complementarities and synergies that increases productivity. In order to achieve these, it pays everyone to help to facilitate exploration, either by sharing the risks in some cooperative way, which takes some of the "sting" out of failure. Once again we must differentiate between an "external game", where total freedom allows wide ranging responses to outside changes, and an "internal game" were the division of labour, internal relations and shared experiences play a role in the survival of the system.

Evolution in human systems is therefore a continual, imperfect learning process, spurred by the difference between expectation and experience, but rarely providing enough information for a complete understanding. Instead of the classical view of science eliminating uncertainty, the new scientific paradigm accepts uncertainty as inevitable. Indeed, if it were not the case, then it would mean that things were pre-ordained, which would be a much harder thing to live with. Evolution is not necessarily progress and neither the future nor the past was preordained. Creativity really exists. It is the motor of change, and the hidden dynamic that underlies the rise and fall of civilisations, peoples and regions. The models discussed here represent a first step towards the development of a scientific approach which captures this truth.

Acknowledgements

The work described in this paper received support from the Open Society Fund, New York, and from the Directorate General VIII of the EEC, under contract 8 946/89.

REFERENCES

Allen P.M., 1976, Darwinian Evolution and a Predator-Prey Ecology, **Bulletin of Mathematical Biology**, Vol 37., pp. 389-405.

Allen P.M., 1978, Dynamiques de Centres Urbains, **Sciences et Techniques**, No.50, pp. 15-18.

Allen P.M., 1982, The Genesis of Structure in Social Systems: The Paradigm of Self-Organization, in **Theory and Explanation in Archaeology**, Academic press, New York.

Allen P.M., 1984, Self-Organization and Evolution in Urban Systems, (**Cities and Regions as Non-linear Decision Systems**, Crosby, ed.) AAAS Selected Symposia, 77, Westview Press, Boulder Colorado.

Allen P.M. Towards a new Science of Complex Systems, **The Science and Praxis of Complexity**, United Nations University Press, Tokyo.

Allen P.M., 1988, Evolution: Why the Whole is greater than the sum of its parts, in **Ecodynamics**, Springer Verlag, Berlin.

Allen P.M., 1990, Why the Future is not what it was, **Futures**, July/August, pp. 555-570.

Allen P.M. and M. Lesser, 1991, **Evolutionary Human Systems: Learning, Ignorance and Subjectivity**, Harwood, Chur, Switzerland.

Allen P.M. and M. Sanglier, 1979, Dynamic Model of growth in a Central Place System, **Geographical Analysis**, Vol 11, No 3.

Allen P.M. and M. Sanglier, 1981, Urban Evolution, Self-Organization and Decision Making, **Environment and Planning A**.

Allen P.M., G. Engelen and M. Sanglier, 1983, Self-Organizing Models in Human Systems, in Synergetics- From Microscopic to Macroscopic Order, **Synergetics Series**, Springer Verlag.

Allen P.M. and J.M. McGlade, 1987, Evolutionary Drive: The Effect of Microscopic Diversity, Error Making and Noise, **Foundations of Physics**, Vol.17, No.7, July.

Allen P.M. and J.M. McGlade, 1989, Optimality, Adequacy and the Evolution of Complexity, in **Structure, Coherence and Chaos in Dynamical Systems**, (Christiansen and Parmentier, eds.), Manchester University Press, Manchester.

Allen P.M. and J.M. McGlade, 1986, Dynamics of Discovery and Exploitation: the Scotian Shelf Fisheries, **Canadian Journal of Fishing and Aquatic Science**, Vol.43, No. 6.

Allen P.M. and J.M. McGlade, 1987, Modelling Complex Human Systems: a Fisheries Example, **European Journal of Operations Research**, Vol. 30, pp. 147-167.

Allen P.M. and J.M. McGlade, 1987, Managing Complexity: a Fisheries Example, Report to the United Nations University, Tokyo.

Eigen M. and P. Schuster, 1979, **The Hypercycle**, Springer, Berlin.

Gould P. and R. White, **Mental Maps**, Penguin Books.

Haken H., 1977, Synergetics, **Synergetics Series**, Springer Verlag, Heidelberg.

Nicolis G. and I. Prigogine, 1977, Self-Organization in Non-Equilibrium Systems, **Wiley Interscience**, New York.

Prigogine I., P.M. Allen and Herman R., 1977, The Evolution of Complexity and the Laws of Nature, **Goals for a Global Society**, (E.Laszlo, ed.), Pergamon Press.

Prigogine I. and I. Stengers, 1987, **Order out of Chaos**, Bantam Books, New York.

Sanglier M. and P.M. Allen, 1989, Evolutionary Models of Urban Systems: an Application to the Belgian Provinces, **Environment and Planning A**, Vol. 21, pp. 477-498.

CHAPTER 9

SPECULATIONS ON

FRACTAL GEOMETRY IN SPATIAL DYNAMICS

Paul Longley and Michael Batty

"Mandelbrot has attracted the attention of scientists on the ubiquity of fractal shapes among natural objects. This was an important and fruitful contribution. What is still missing in general is an understanding of how fractal shapes arise."

David Ruelle (1991) Chance and Chaos, Princeton University Press, Princeton, NJ, p. 178

9.1 THE DEVELOPMENT OF FRACTAL GEOMETRY

The most exciting science always stands on the frontier between the known and the unknown, and in the last decade in the effort to push our understanding of real and natural systems with all their imperfections to the limit of our analytical powers, fractals and their geometry have attracted very widespread attention. The idea that the geometry of Euclid is but a special case of a more general geometry in which the irregular and the fractional rather than the regular and the integral represent the norm, is appealing in its intuitive sense, and thus fractal geometry has spread like wildfire throughout the sciences. Applications of course abound. Although science is ordered in that its theories deal with phenomena which are represented systematically and with regularity, its practice through applications is anything but for real systems seem to be at the other end of the spectrum. Everywhere we look from the stock market to the form of river systems, from crystal growth to the way we pack circuits onto a silicon chip, shapes which can only be described in fractional or fractal dimension seem to exist.

Yet as David Ruelle (1991) who we have quoted at the outset of this essay so cogently remarks, how fractals arise, how they emerge and evolve, change and grow has hardly been broached as yet for many systems. Our ability to describe and simulate their shape through various shortcuts, however ingenious, is nowhere matched by any ability to explain how such systems develop through time as well as space. So far little has been achieved in developing a unified fractal geometry applicable to consistent and integrated theories of space-time. In fact, what we will do in this essay is present ideas relating to our abilities to use fractal geometry to describe static spatial systems although in the spirit of this book we will speculate on how these ideas might be extended to truly dynamic spatial models which build on theories of nonlinear dynamics whose form can also be presented in terms of fractal geometry.

Fractal geometry is the geometry of irregular self-similar shapes that exist across many scales. Such shapes include both regular 'mathematical' shapes, as well as irregular 'natural' shapes, such as coastlines and trees, which manifest abrupt changes in continuity. Since their popularization by Mandelbrot in the late 1970s (Mandelbrot 1982), fractal concepts have found wide application in a range of physical sciences and, more recently, in human systems (see Feder, 1988 for a review). The geometry as originally developed includes the geometry of real Euclidean space and its generalization to n-dimensional mathematical space as a special case. It is, in essence, a geometry which relaxes the constraint of integral dimension and permits measurement and simulation across a continuum of fractional (hence fractal) dimensions. It has been applied to temporally dynamic systems insofar as changes in scale over time have been shown to be fractal but the connections between spatial and temporal scales are tenuous. In addition to briefly reviewing this research, the purpose of this paper is to investigate how fractal geometry might be used to measure and simulate changes in scale across space, and the ways in which such developments might have implications for temporal dynamics.

Over the same period, there have also been dramatic developments in the analysis of nonlinear systems, with particular emphasis upon the temporal irregularities and discontinuities which characterize them. Early developments in catastrophe theory and bifurcation were made during the 1970s (Thom, 1975; 1983; Nicolis and Prigogine, 1989), and during the 1980s, various applications to city and regional systems were made (Wilson, 1981; and Allen, 1991). These developments are now viewed within the broader context of nonlinear dynamic models with chaotic equilibria, as developed extensively during the last decade (Holden, 1986; Ruelle, 1991; Schroeder, 1991). Such chaotic dynamic models are characterized by equilibrium attractors which display fractal geometric structure in their phase space. Our own research concerns the application of fractals in a spatial context, although more generally fractals hold the enticing prospect of unifying our conceptions of space and time through the idea of self-similarity across different spatial and temporal scales.

In one sense, chaos and its foundations in nonlinear dynamics has become much more popular than fractal geometry (Gleick, 1988). Real systems are clearly more random and less explicable than those that one learns about in economics and physics and the fact that the weather and the stock market, everyday topics of conversation, could be explicable as chaotic systems has clearly wetted the popular imagination. The pioneering work of Lorenz (1963) cast the conception of weather as a chaotic system, in which large scale outcomes are conceived as being extremely sensitive to initial conditions: the geometry of the different trajectories from minutely varying initial conditions exhibits self-similarity and the system is, as such, a chaotic fractal. Similarly the ecological dynamics of predator-prey relations show extreme sensitivity to initial conditions, insofar as oscillations depend crucially upon initial parameter values. In a world with newly found interests in questions of environment and the precarious balance of resources globally, ideas that we might be dealing with intrinsically chaotic systems which we might understand but be unable to control are appealing.

Our last parallel development which we consider important has been the increasingly routine innovation of computer graphics into science and social science computing. Although the geometry of fractal shapes such as the Koch and Peano curves has been recognized for generations, it is only through harnessing the recursive power of computer graphics that generation and display of such shapes has become commonplace. Introduction of randomness (most usually based upon a model of Brownian Motion) into the recursive process enables exploration of a variety of geometrical forms which can come to assume real world visual characteristics. The property of self-affinity has also been successfully used to generate plausible real-world fractal shapes (Barnsley, 1988). Two key icons have become important in the popularization of fractals. First, the realistic modelling of terrain through the process of fractal rendering has led to the production of a number of well-known fractal images (e.g. Fournier et al, 1982) and movie sequences, providing testimony to the power and practical application of fractal ideas. The second is the Mandelbrot Set in which space is mathematical and real time is represented in terms of the mathematical dynamics of solution of a quadratic equation in the complex plane. These widely differing examples which nevertheless manifest the same geometry provide an immediate sense of the breadth and diversity of scientific investigation of the form and origins of fractal shapes.

Fractal geometry thus enables us to deal with a new sort of regularity which defies measurement using conventional geometries. This enables us to gain a new and enhanced understanding of how idealized systems can be mapped into real systems, and holds the prospect of our reconciling idealized or normative models with positive forms. In the next Section, we will address the ways in which these ideas have begun to find a niche in the traditional spatial disciplines of geography and regional science.

9.2 CHAOS AND FRACTALS IN GEOGRAPHY AND REGIONAL SCIENCE

During the 1950s and 1960s, geographers and regional scientists were responsible for the development of comparatively simple spatial and temporal models of city and regional systems based upon the ideals of Euclidean geometry and linear dynamics. Such models involved the development and extension of Central Place Theory using regular hexagonal lattices, through to the development of various econometric space-time models (e.g. Paelinck and Klaassen, 1979). Since the 1970s, such models have been made more realistic in spatial and temporal terms through the use of nonlinear geometries and nonlinear dynamics. It is in this particular spirit that fractal geometry has been of such importance and offers such potential. In broad terms, we can consider these developments under three headings: population models; spatial hierarchies and market area analysis; and the development of discrete spatial systems using dynamic models. We will discuss each of these in turn, before reviewing our own work within this broader context.

Population modelling has developed through its association with ecological dynamics and the notion of bifurcations in equilibrium trajectories (as embodied in Volterra-Lotka models: see Peitgen and Richter, 1986). Simple applications of these models were first espoused by Dendrinos and Mullally (1985) amongst others. Generalizations of more complex Volterra Lotka models (Dendrinos and Sonis, 1990) based upon more intricate predator-prey relations and arms races between different groups lead to chaotic equilibria whose phase space attractors can be interpreted as fractal geometries. These types of models have been used to model the development of city systems in terms of spatial equilibria, particularly by Wilson (1981), and Allen (1993). This work has explored ideas from nonlinear dynamics and has begun to broach ideas from chaos, but so far, has not explicitly presented these models or systems in terms of their fractal geometry.

The second main area of development is in terms of spatial hierarchies and regular tessellations of market areas (Arlinghaus, 1985). Rank size and central place hierarchies have been shown to be entirely consistent with fractal concepts that are based upon scaling in that the geometry of such hierarchies is self-similar across a range of scales. There are strong links here with network geometry (Haggett and Chorley, 1969) and the physical geography of river systems, since both are intrinsically fractal (Batty, 1992a, 1992b). Although there has been some speculation as to how existing theories and models might be reinterpreted, no comprehensive and detailed research agenda has not yet been produced.

Our own work is, however, based upon two developments of these ideas. First, we have been concerned with measuring the properties of urban and regional city systems using fractal geometry. In this work we have been concerned with the structure and irregularity of land use and urban development in space. We have concentrated upon measurement of the shape and morphology of land use and development through interpreting boundaries as fractal geometries. Such measurements have been used, on the one hand, to identify the structure and character of cartographic lines and, on the other, to program the structured irregularity of simulated lines. In spirit this work is similar to rendering based on the sort of visual forms which are found in terrain and in remotely sensed images. Our second related area of research has been in modelling the growth of cities and their resulting forms. Our various simulations have been based upon models of constrained diffusion of growth about seed sites, including diffusion-limited aggregation (DLA) and dielectric breakdown (DB). The structures produced by these models exhibit strong visual similarity with network structures and can be seen to develop in a similar manner. There is a great deal of research in diffusion in geography, beginning 40 years or so ago in the work of Hagerstrand (1953), much of which has not yet been related to this area. Moreover, our various kinds of DLA and DB simulations might be related to more mainstream space-time dynamics: specifically, there exists the prospect of developing space-time models in which fractal geometry might be used to model both space as well as time.

The third area involves the development of discrete spatial systems through simulating spatial dynamics rather than devising models based upon the classical continuous mathematics of differential equations and the like. Such cellular automata are developed by invoking simple rules based on spatial transitions, normally between adjacent cells of a matrix, and were first applied in geography in the mid-1970s by Tobler (1979) and subsequently by Couclelis (1985) and White and Engelen (1993). Like fractal geometry and nonlinear dynamic models, the processes governing the growth of such cellular automata have been understood for over 30 years: the approach has also received a considerable stimulus from the widespread use of digital computation and, latterly, computer graphics (Wolfram, 1984). This is an area of research which has experienced very rapid growth in recent years, and has recently become a popular topic through its relation to theories of human evolution (Allen, 1993; Levy, 1992; Holland, 1992). It is not hard to see it also has far-reaching implications for geography and regional science, many of which are likely to be developed in some depth over the next decade.

9.3   APPLICATIONS TO SPATIAL DYNAMICS

Thus far in this paper we have sketched out the broad brush agenda for using concepts of fractals and chaos in geography and regional science research. In order to avoid the danger of being typecast as agenda- setters rather than 'coal face' researchers, we will now describe our own contributions to this broader effort. The first broad theme of our research has concerned the use of fractally-rendered partitions of space to present large-scale spatial forecasts based upon the results of urban modelling work. It was with some disquiet that we observed (Batty and Longley, 1986) that urban models all too often assumed space away and that the inductive process of model-building was rarely complemented with deductive use of model parameters to drive large-scale spatial forecasts. Even in the comparatively few studies in which spatial forecasts were made, predictions were restricted to coarse zonal aggregations, rather than the smaller geographical building blocks that in reality comprise the urban fabric.

Figure 1 presents two forecasts of the occurrence of four dwelling types in Greater London, based upon deterministic interpretations of two slightly different discrete choice model specifications. The simulation space of each map is partitioned up with the use of a fractal generator which we deemed characteristic of the signatures of urban land use parcels in the Greater London area. Although not accurate in an absolutist geographical sense, the simulations nevertheless generate considerable visual realism and provide a useful adjunct to established numerical comparative statistics and exploratory diagnostics. At one level, therefore, this research enables models to communicate more effectively. Behind this, however, lies the notion that if we view the world as fractal, we can begin to explore the link between models which work at a fairly abstract spatial scale, but can then transfer the results to a more meaningful scale. It also requires us to concentrate thinking about contiguity and adjacency, which in turn stimulates thinking about the relationship between spatial form and process.

The second strand to our own research begins to develop these links between form and process, with the goal of making spatial forecasts more than just an appendage to the model-building process. We have begun with a reappraisal of the classic Von Thünen model of agricultural land use, in which different land uses are located at different distances from the market based on their profitability, and its more elaborate extensions to urban location theory which seek to explain why rents and population densities fall with increasing distance from city centers. The ideal geometric patterns which form the cornerstones of contemporary human geography represent ideal types but as every geography student knows, real regions and cities do not correspond to these idealized patterns. Physical constraints on land development, the existence of competing centers, the idiosyncrasies of politics and history, human inertia, and a host of other imperfections in the market combine to make real urban and regional patterns somewhat different from these ideal types. It is here perhaps that fractal geometry is coming to play its most fundamental part. Using a randomized version of Von Thunen's model to locate commercial/industrial land use, residential housing and open space, we can use fractal geometry to introduce some spatial realism into such a simulation. Figure 2 shows such a simulation for a large city with approximately the same dimensions as London while in Figure 3 we have shown the actual distribution of the population of London taken from the 1981 Census and rendered using a new technique of computer display developed by Ian Bracken at Cardiff University (Martin and Bracken 1991). Both the simulated and real maps of London show patterns which are clearly fractal in that the location and form of development is statistically self-similar across a range of scales.

A somewhat clearer picture of this fractal form can be illustrated through the changing boundary of a city as it grows through time. In Figure 4 we show the boundary of the urban growth of Cardiff at intervals from 1886 to 1947. These boundaries are in some respects like coastlines (Longley and Batty, 1989a; 1989b). We can recognize self similarity across different scales and demonstrate this through the fractal dimensions of these boundaries which have been calculated as 1.24, 1.18, 1.17 and 1.29 for the years 1886, 1901, 1922 and 1949. But perhaps the most characteristic form displayed by these pictures of cities involves the way in which their urban growth seems to reach out from their centers in tentacles of development. This phenomenon is clearly seen in the maps of London and Cardiff and it manifests itself best in the growth of 'ribbon' development along major transport routes such as roads and railways which emanate from the city's centre. In fact, we have begun to model this as a diffusion- limited aggregation (DLA) process based upon the notions of irreversible growth and transition of structures far-from-equilibrium (Batty et al, 1989).

To simulate the growth of the city itself, we can adopt this widely used process of 'growing fractals' from single seeds and we will take a minute to describe how this is done. If you plant a seed somewhere near the centre of a two-dimensional region, and then launch a particle which we will assume is a unit of urban development at some distance from this centre, you can begin a random walk with the particle which will either ultimately wander off the map crossing its edge or will eventually come adjacent to the seed site and 'stick'. Once either event occurs another particle is launched, it begins its walk and eventually it either wanders off the plane or reaches a particle which has already stuck and 'sticks'. If this process continues, then what will happen is that tentacles of development will grow out from the seed site, this dendritic pattern being reinforced as the process continues.

What is perhaps surprising about this is the fact that a tree-like structure resembling transport routes in a large city develops rather than a circular spread of particles around the central site. Basically this is because once a particle sticks, this reinforces the direction of growth thus leading to tentacles between which the location of new particles becomes increasingly less likely. Such a pattern is shown in Figure 5 where the tree-like form is clearly similar to the way a western industrial city might develop although unlike real cities, the units of development are all connected to one another. The whole two-dimensional region is not filled by this development but the development is a lot more than a one-dimensional line of growth. In fact the fractal dimension of such a structure is about 1.7 and similar processes of growth occur throughout physics, especially where there are transitions in the state of the system as in the crystallization of particles in solution, the breakdown of electricity such as lightening, the permeation of oil slicks in water, and such like. The form of these patterns is clearly self-similar in that the tree-like structures branch in similar ways at whatever scale the structure is examined.

Patterns like these occur in cities especially in transport networks. Benguigui and Daoud (1991) have shown that these sorts of patterns are close to transport networks such as the Paris railway system while we can also use the model to simulate urban growth. In Figure 6 we show how such a process can be used to mimic the growth of the town of Cardiff which is constrained by the natural barriers of the Severn Estuary and rivers within the town, and it is clear when we examine the growth of London as in Figure 3 that such tree-like patterns exists across many scales. In fact in all the pictures we have shown here we have not shown actual scales thus illustrating the basic notion that fractals manifest similar forms across many scales. We should also note that the forms of city which we can simulate as growing fractals not only produce realistic patterns but also generate densities of population and employment and intensities of land uses which are quite consistent with well-established examples and ideas.

9.4. CONCLUSIONS AND FUTURE RESEARCH

What is clear is that the application of concepts from fractal geometry and chaos offer us enticing new ways of both conceptualizing and visualizing urban and regional phenomena. In this work, measurement prescribes analysis, and we have made a start in diagnosing the structure and character of both real world and modelled urban forms. As our expertise in such analysis develops, we will begin to relate the geometry of urban form to the density of settlement. Recently one of us (Batty and Kim, 1992) has suggested that densities can be interpreted as scaling functions, specifically that the inverse power function can be interpreted as a fractal dimension. If this dimension were to lie in the region of 1.71, this might be taken to be indicative of a universal scaling function which describes the general morphology of cities. These ideas can be easily extended to spatial phenomena in general. If the key parameter of a spatial interaction model based on the inverse power law is interpreted as a fractal dimension, a consistent link between the way development fills space and this value can be developed: empirical evidence suggests that the estimated parameter value of such models is less than two, and our speculation at this point would be that it may lie close to the space filling parameter value of 1.71.

In fact we might go so far as to suggest that power functions are more appropriate measures of distribution in spatial systems than are exponential for such functions are scaling and as such are consistent with the sorts of self-similarities we find in urban development at different scales. If this argument is accepted, it might suggest that the parameter value of a spatial interaction model should be between 1 and 2, probably nearer 2 than 1, say 1.7, and that the density parameter associated with this model should thus be equal to $2 - 1.7 = 0.3$. Much more work is required in this spirit before such universals are even worthy of consideration but the missing link between morphology and spatial interaction which has plagued the subject of geography for so long may well be found in such developments.

Finally, the development of extensive surface models of population phenomena as described by Martin and Bracken (1991) considerably broadens the remit of empirical analysis of the size, shape and dimension of urban settlements. There exists the prospect

(Longley et al, 1992) of classifying cities and city systems using fractal dimensions, and consequent evaluation of the importance of regional and urban planning policy in the delimitation of form. Linking such data and applications to the enormous data base available from remotely sensed data also has great potential in the application and testing of these ideas in the near future. These and other topics are part of our own research agenda but the sorts of speculation which we began this paper with are worth spelling out in a little more detail. For fifty years or more, spatial analysts have sought to criticize economists and other social scientists for their ignorance of the effect of space while at the same time finding it immensely difficult to develop consistent treatments of time themselves. The conditions now seem appropriate, however, for the development of much more consistent space-time models using such ideas as those that are appearing in fractals and chaos theory, and it is possible that computer models in which morphologies systematically evolve and change are likely to be developed in this spirit in the next decade, linking the ideas of this paper to many of those presented in other parts of this book.

ACKNOWLEDGEMENT

The authors are grateful to Ian Bracken (University of Cardiff) for permission to reproduce the image shown in Figure 3.

REFERENCES

Allen, P., 1991, Spatial Models of Evolutionary Systems, (D. Pumain, ed.), **Spatial Analysis and Population Dynamics**, John Libby, Montrouge, France, pp. 147-160.

Allen, P., 1993, Spatial Evolution in Complex Systems, this volume.

Arlinghaus, S.L., 1985, Fractals Take a Central Place, **Geographiska Annaler**, 67B, pp. 83-88.

Barnsley, M., 1988, **Fractals Everywhere**, Academic Press, London.

Batty, M. and P.A. Longley, 1986, The Fractal Simulation of Urban Structure, **Environment and Planning A**, 18, pp. 1143-1179.

Batty, M., P. Longley, and A.S. Fotheringham, 1989, Urban Growth and Form: Scaling, Fractal Geometry and Diffusion-limited Aggregation, **Environment and Planning A**, 21, pp. 1447-1472.

Batty, M., 1992a, The Fractal Nature of Geography, **Geographical Magazine**, 64, pp. 32-36.

Batty, M., 1992b, A Random Walk Along the Urban Boundary, **Geographical Magazine, 64**, pp. 34-36.

Batty, M. and K-S Kim, 1992, Form Follows Function: Reformulating Urban Population Density Functions, **Urban Studies**, in press.

Benguigui, L. and M. Daoud, 1991, Is the Suburban Railway System a Fractal?, **Geographical Analysis**, 23, pp. 362-368.

Couclelis, H., 1985, Cellular Worlds: a Framework for Modeling Micro-macro Dynamics, **Environment and Planning A**, 17, pp. 585-596.

Dendrinos, D.S. and H. Mullally, 1985, **Urban Evolution: Studies in the Mathematical Ecology of Cities**, Oxford University Press, Oxford, UK.

Dendrinos, D.S. and M. Sonis, 1990, **Chaos and Socio-Spatial Dynamics**, Springer Verlag, New York.

Feder, J., 1988, **Fractals**, Plenum, New York.

Fournier, A., D. Fussell, and L. Carpenter, 1982, Computer Rendering of Stochastic Models, **Communications of the ACM**, 25, pp. 371-384.

Gleick, J., 1988, **Chaos: Making A New Science**, Heinemann, London.

Hagerstrand, T., 1953, **Innovation Diffusion as a Spatial Process**, University of Chicago Press, Chicago, Illinois, (translated 1967).

Haggett, P. and Chorley, R.J., 1969, **Network Analysis in Geography**, Edward Arnold, London.

Holden, A.V., (ed.), 1986, **Chaos**, Manchester University Press, Manchester, UK.

Holland, J. H., 1992, **Adaptation in Natural and Artificial Systems**, MIT Press, Cambridge, Massachusetts.

Levy, S., 1992, **Artificial Life: The Quest for a New Foundation**, Pantheon, New York.

Longley, P. and M. Batty, 1989a, Fractal Measurement and Cartographic Line Generalisation, **Computers and Geosciences**, 15, 2, pp. 167-183.

Longley, P. and M. Batty, 1989b, On the Fmeasurement of Geographical Boundaries, **Geographical Analysis**, 21, 1, pp. 437-452.

Longley, P., M. Batty, J. Shepherd, and G. Sadler, 1992, Do Green Belts Change the Shape of Urban Areas? A Preliminary Analysis of the Settlement Geography of South East England, **Regional Studies**, 26, in press.

Lorenz, E.N., 1963, Deterministic Non-periodic Flows', **Journal of Atmospheric Sciences**, 20, pp. 130-141.

Mandelbrot, B.B., 1982, **The Fractal Geometry of Nature**, Freeman, New York.

Martin, D.J. and I. Bracken, 1991, Techniques for Modelling Population-related Raster Databases, **Environment and Planning A**, 23, 1069-1075.

Nicolis, G. and I. Prigogine, 1989, **Exploring Complexity: An Introduction**, Freeman, New York.

Paelinck, J.H.P. and L.H. Klaassen, 1979, **Spatial Econometrics**, Saxon House, Farnborough, UK.

Peitgen, H-O. and P.H. Richter, 1986, **The Beauty of Fractals: Images of Complex Dynamical Systems**, Springer Verlag, Berlin.

Ruelle, D., 1991, **Chance and Chaos**, Princeton University Press, Princeton, New Jersey.

Schroeder, M., 1991, **Fractals, Chaos, Power Laws**, W.H. Freeman, New York.

Thom, R., 1975, **Structural Stability and Morphogenesis: An Outline of a General Theory of Models**, Benjamin, Reading, Massachusetts.

Thom, R., 1983, **Mathematical Models of Morphogenesis**, Ellis Horwood, Chichester, UK.

Tobler, W., 1979, Cellular Geography, (S. Gale and G. Olsson, eds.), **Philosophy in Geography**, Reidel, Dordrecht, Holland, pp. 279-386.

White, R., 1992, Cellular Dynamics and GIS: Modelling Spatial Complexity, a paper presented at the NCGIA Initiative 14: GIS and Spatial Analysis, San Diego, California (available from NCGIA, SUNY-Buffalo, New York).

White, R. and G. Engelen, 1993, Complex Dynamics and Fractal Urban Forms, this volume.

Wilson, A.G., 1981, **Catastrophe Theory and Bifurcation: Applications to Urban and Regional Systems**, Croom Helm, London.

Wolfram, S., 1984, Cellular Automata as Models of Complexity, **Physica D**, 10, pp. 1-35.

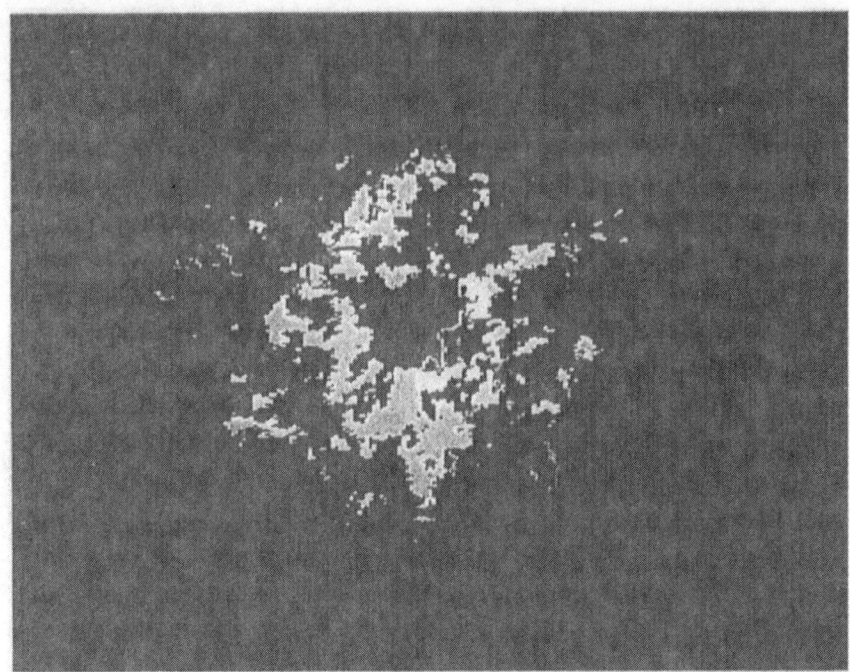

Figure 1. Visualizing the Morphology of Dwelling Type Generated from a Discrete Choice
Model of Housing Tenure in Greater London.

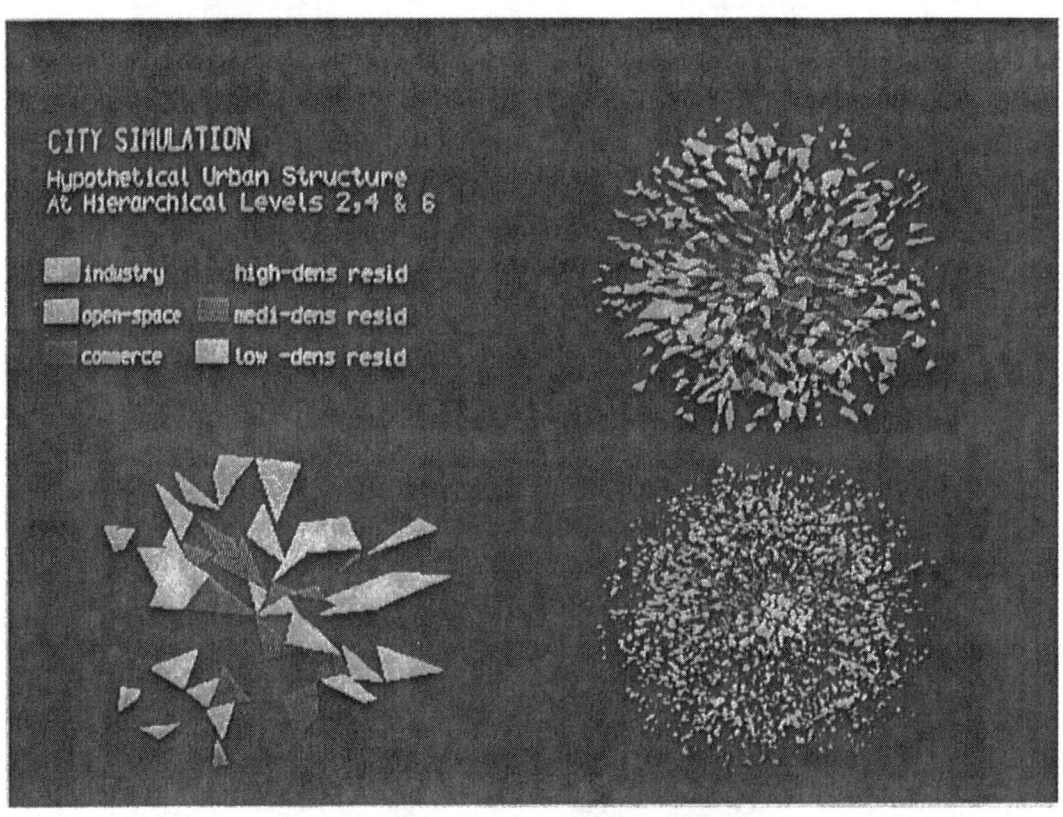

Figure 2. A Simulation of Large-Scale Urban Development Using a Fractal Rendition of Predictions from a Von Thünen Model.

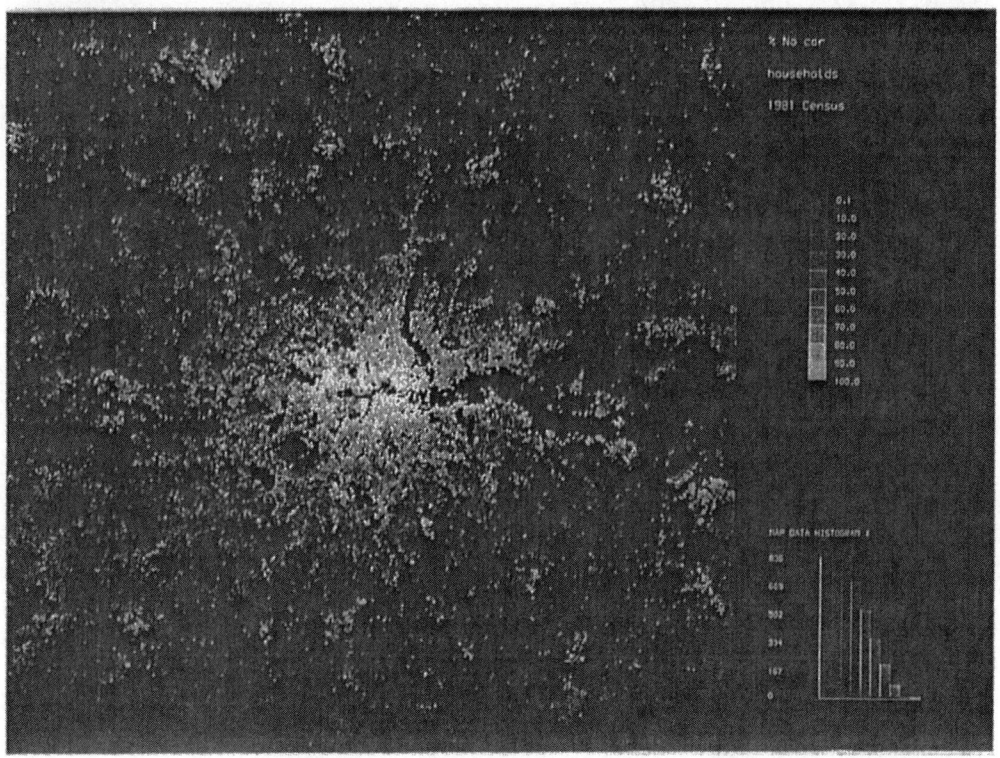

Figure 3. The Morphology of Greater London from the 1981 Population Census Using Bracken's Method.

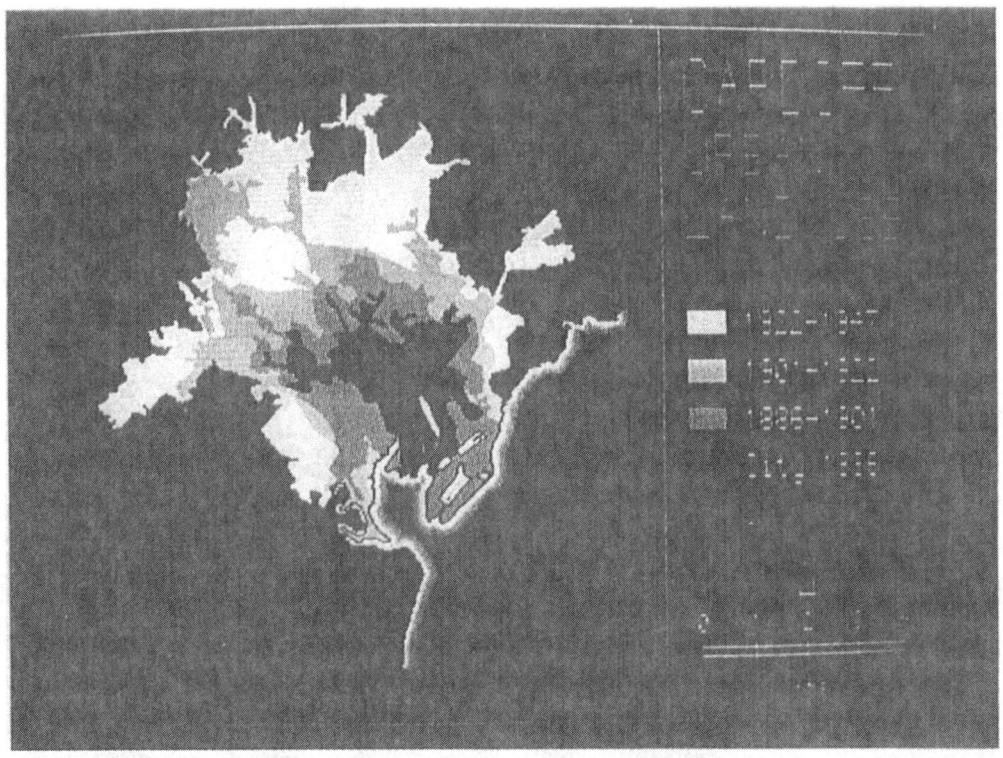

Figure 4. The Growth of Urban Cardiff from 1886 to 1947.

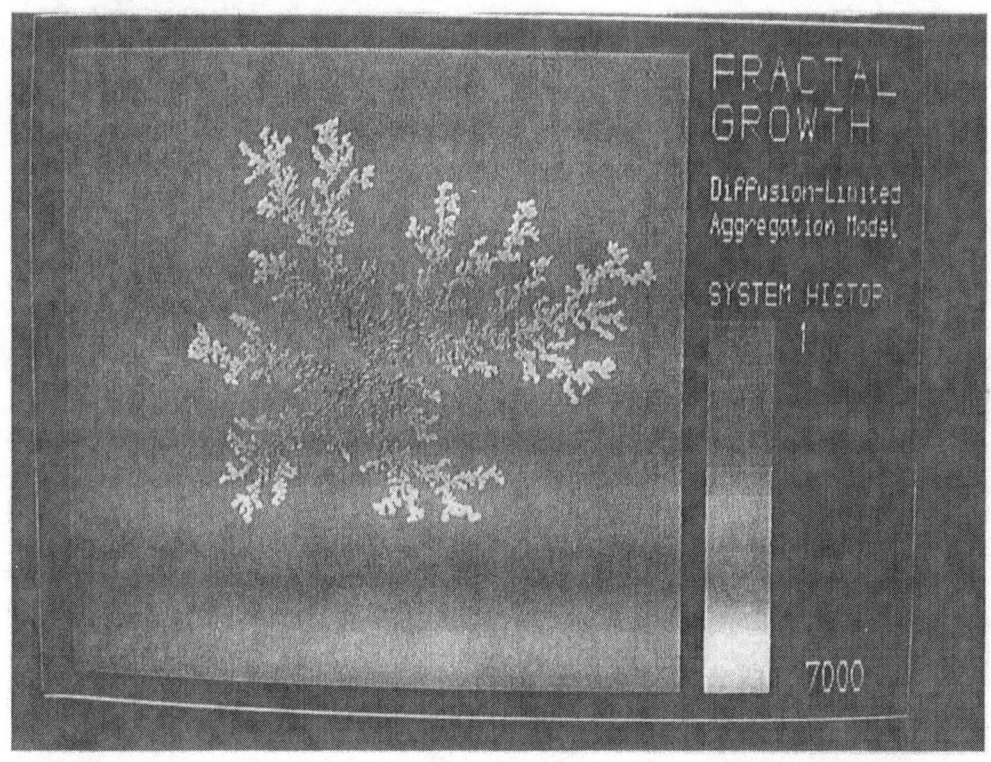

Figure 5. Diffusion-Limited Aggregation About a Central Seed Site.

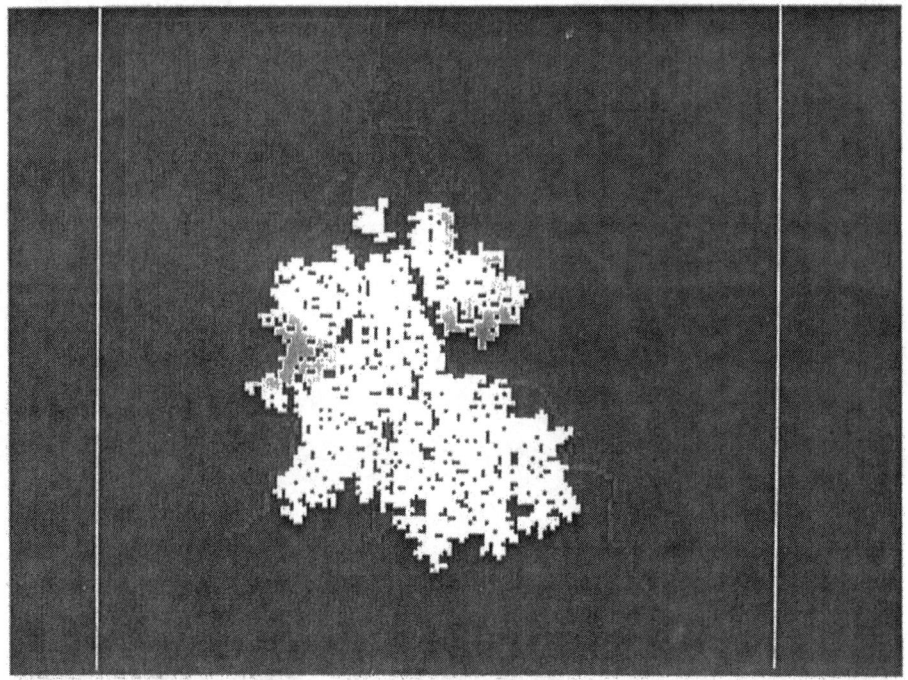
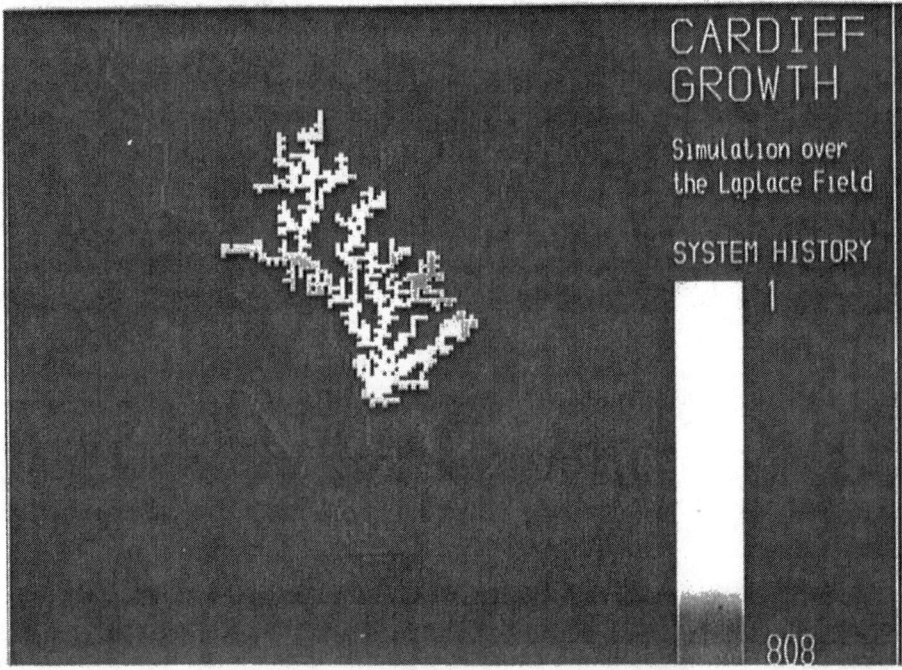

Figure 6. Simulating the Growth of Cardiff Using the Dielectric Breakdown Model.

CHAPTER 10
COMPLEX DYNAMICS AND FRACTAL URBAN FORM
Roger White and Guy Engelen

10.1 INTRODUCTION

A first-time visitor to Paris or Los Angeles, or any other city, is acutely aware that despite some relatively predictable features, cities are highly diverse and elaborately complicated. In terms of land use, residential, industrial, and commercial areas are interspersed with each other at various scales and in a variety of shapes. All cities, with the exception of some planned communities, have this intricate spatial quality, so it would seem to be important. Indeed, recent results in abstract systems theory support this conclusion. Work by Kauffman (1990), Langton (1990) and others suggests that only systems which are sufficiently complex - specifically, only systems that are nearly chaotic - have the ability to evolve. Since evolvability is clearly an essential characteristic of all living systems, including social systems, these results imply that cities **must** be inherently complex.

Traditional approaches to the problem of urban structure are essentially either historical or economic in nature, but almost never both. Urban historians, tracing sequences of individual events, and insofar as possible revealing the motives of the actors involved in these events, frequently provide very satisfying insights into the development of particular cities. But little is revealed about general principles that might explain the development of urban systems. Urban economists, on the other hand, typically concentrate on a search for organizing principles, but take an a-temporal approach. Urban structure is explained as the outcome of individual decisions to maximize utility or profit subject to relevant constraints such as income or availability of land. Typical models, such as Alonso - type land use models, yield simple, radially symmetric, geometrical predictions for urban form.

At the level of the individual actors involved in the system, the description of economic behaviour may be quite reasonable, since a city is, in many ways, clearly the product of such decisions. But at the aggregate level these models fail to represent a temporal process; rather they describe a static general equilibrium in which every individual is at a constrained optimum. In this respect the approach is completely antithetical to the nature of the phenomenon under investigation, since real cities develop in time, and being constrained by their past, are unlikely ever to reach an optimum state. Indeed, since cities are evolving, dissipative, far-from-equilibrium structures, a general equilibrium approach is clearly inappropriate.

Dynamic approaches, on the other hand (eg. Allen, 1983; Allen and Sanglier, 1979; Allen et al, 1984; Engelen, 1988; White, 1977, 1978, 1984; Wilson, 1978), are more appropriate in that they do not depend on an assumption of equilibrium. Furthermore, the spatial patterns emerge from a relatively complex non-linear dynamical process and are thus themselves more complex and hence realistic. Models specified to represent actual cities (eg. Pumain et al, 1987), have given reasonable approximations to actual urban form.

However, while these models may exhibit quite complex dynamics, the degree of spatial complexity which they can represent is limited. In these models, in general every region interacts with every other one. Computational requirements consequently increase very rapidly as the number of regions is increased, so that in practice the models become impractical when a high degree of spatial disaggregation is desired. In contrast, in the cellular approach developed in this paper, it is natural and feasible to use a very large number of regions - i.e. cells - and thus to achieve great spatial detail. For example, a model of a city 20 km in diameter developed on a 500 x 500 array would contain 250,000 regions, and would have a spatial resolution of 40 meters - the scale of an individual block.

Furthermore, aside from the computational constraints, there is evidence that the relationships described by the equations in the traditional dynamic models do not hold over short distances. At very local scales, other effects become important determinants of the spatial organization of the system. Thus a typical dynamic model predicts for each time period a certain level of residential, commercial, office and industrial activity for each zone, but says nothing about the arrangement of these activities within that zone. Yet the quality of their spatial arrangement may be a major determinant of the future attractiveness of the region. And problems at this local, intra-regional level are among the most important for both residents and planners.

Perhaps the best way to understand the complex order hidden in spatial detail is to know the rules or processes that generate the complexity. Fortunately, it seems to be often the case that much of the complexity is produced by relatively simple processes. Thus a useful approach is to postulate likely behavioural rules and observe the spatial structures that result from them. Cellular automata are ideally suited for this task, since they combine explicitly spatial processes with a high degree of spatial resolution. The focus of this paper is the development of a cellular model of urban land use. The model is used to investigate theoretical questions of urban form. Empirical control is provided by data from US cities.

10.2   A CELLULAR MODEL OF URBAN FORM

A cellular automaton consists of an array of cells which may be in any one of several states. At each iteration, the state of each cell either remains the same or changes to another state depending on the states of the neighbouring cells. Models differ from each other essentially in the nature of the transformation rules that relate the state of a cell to that of its neighbours. In one of the simplest but most widely studied models, for example, the Game of Life, cells are in one of two states: alive and dead. A live cell stays alive only if either two or three of the adjacent cells are alive; otherwise, it dies. And a dead cell will come to life if it has exactly three live neighbours.

Many cellular automata exhibit self-organized criticality--that is, their dissipative nature moves them continually from one metastable state to another, but all the metastable states of a given automaton are qualitatively similar and are characterized by the same fractal dimensions.

Cellular automata are currently being used in a variety of disciplines to investigate fundamental questions concerning the origin and evolution of structure (Bak and Chen,

1989; Bak et al, 1989; Grassberger, 1991; Markus and Hess, 1990; Phipps, 1989). Thus Bak and Chen (1989) show that a few simple models can give insights into a wide variety of phenomena. Their Forest Fire model, for example (but see also Grassberger, 1991), demonstrates the mechanism by which a spatially uniform energy input is dissipated on a consequent fractal structure. The mechanism is relevant not only to percolation processes, but also to the spread of epidemics or chemical reactions, and to the appearance of turbulence. These models are, however, by their very generality, not realistic descriptions of any particular phenomenon.

Tobler (1979) was perhaps the first to propose the application of cellular techniques in geographical modelling. Couclelis (1985, 1988,1989) followed with a series of essentially conceptual models based on the principles of cellular automata. Although only one of these models is implemented, used essentially as thought experiments, they provide valuable insight into the nature of spatial process and structure. Finally, Hillier and Hanson (1984) have used iterated cellular techniques to model very local spatial structures - ranging in scale from the arrangement of buildings in small villages to the layout of rooms in a house. They succeed in showing that simple rules applied locally can generate forms that closely resemble the real structures.

Another class of models closely related to cellular automata - diffusion limited aggregation models - is being used to model clustered forms, including cities. These models also typically yield fractal forms, and while the process modelled does not seem to correspond to any actual urban growth process, the studies are none the less interesting and contain much useful discussion. (For urban applications, see Batty et al, 1989, Fotheringham et al, 1989, and Batty, 1991a, 1991b.)

The cellular automaton developed here is designed primarily to investigate basic principles of urban spatial form; therefore it is kept as simple as possible. Nevertheless, since it must give a reasonable representation of a city, it is significantly more complicated and more specific than the highly generic models like Forest Fire and Game of Life. It is essentially a model of urban land use, with cell states representing type of land use, for example vacant, commercial, industrial, or housing; and the cellular city grows and evolves as cells are converted from one state to another. The model is specified as follows.

1. Each cell can have one of four states, each representing a land use: vacant, housing, industrial, or commercial. Furthermore, the states are hierarchical in the order listed: a cell can only change to a higher state. Thus, for example, a vacant cell can change to any other state, but an industrial cell can only change to commercial; commercial cells cannot change state. Since occupied cells cannot become vacant, the city can only grow. This unidirectional change of state is imposed for computational reasons, and is being removed in a subsequent version of the model. While it is not an inherently desirable feature, it has not prevented useful results from being obtained.

2. The net number of cells N_i to be converted to each non-vacant state i at each iteration is determined exogenously to the evolution of the cellular automaton. The net numbers to be converted to each state in the first iteration are supplied as input data, and the numbers to be converted in each subsequent iteration are calculated by applying a specified growth rate (eg. 5%) to the figures for the previous iteration. The gross numbers to be converted are in general higher, since, for example, some housing cells will be converted to industrial or commercial use, and so additional cells must be converted from vacant to housing to compensate for these transitions.

Specifying the growth rate exogenously seems entirely reasonable, since the growth of a city would normally depend on factors such as the city's role in the national economy rather than on its internal spatial structure. It is less obvious that the proportions of land occupied by the three activities should be exogenous, but again, it can be argued that the amounts of industrial and commercial activity, relative to housing, in a city depend largely on the city's role in the regional, national, and international economies.

3. The fate of a cell at each iteration depends on the states of the cells within a neighbourhood surrounding it, where the neighbourhood may have a radius of up to six units (where a unit is one cell width). Since the array of cells is regular, each cell within the neighbourhood falls within one of 18 discrete distance categories.

For each non-commercial cell, transition potentials must be calculated for all allowed transitions, that is for transitions to all higher states. Specifically, the preliminary or deterministic transition potentials for a cell are calculated as weighted sums:

where P_{hj} is the transition potential from state h to state j,
m_{kd} = the weighting parameter applied to cells with state k in distance zone d
i = index of cells within a given distance zone
I_{id} = 1 if the state of cell i = k; otherwise, I_{id} = 0.

Thus, cells within the neighbourhood are in general weighted differently depending on their state and also depending on their distance from the cell for which the neighbourhood is defined. Since different parameters can be specified for different distance zones, it is possible to build in weighting functions that have distance decay properties similar to those of traditional spatial interaction equations.

4. The deterministic transition potentials are next subjected to a stochastic perturbation, in which each potential is multiplied by a stochastic disturbance term, s:

$$s = 1 + (-\ln(rand))^n \qquad (10.2)$$

where $0 < rand < 1$ is a uniform random variate and n is a parameter. Thus s ranges from unity to infinity, and n determines the relative importance of the stochastic disturbance. Since from equation (10.1) the deterministic transition potential $P_{hj} >= 1$, every cell in the array has a non-zero chance of transition.

5. To select the cells to be transformed to each function at each iteration, the potentials calculated for each cell for transition to a particular state are ranked. Beginning with potentials for transformation to the top state, commerce, the N_c cells with the highest potentials are identified. If some of these cells also appear among the N_i cells with the highest potentials for transformation to industry, or the N_h cells with the highest housing

potentials, then a series of stochastic tests is made to determine which of them are retained for conversion to commerce, rather than converted to one of the other functions. The probability that a cell will be retained for conversion to commerce is specified as a parameter. Once the N_c cells have been selected for transition to commerce, the procedure is repeated for conversions to industry and housing. The procedure just described is necessary because the probabilistic transition potentials are scaled arbitrarily, depending on the values of the parameters m. Thus the potentials for transformation to different states cannot be compared directly. Only comparisons of potentials for transition to the same state are valid.

6. The model is run for as many iterations as possible before sizable clusters of urban activity impinge on the edge of the array and boundary effects become important. The higher the growth rate specified, and the higher the exponent n in the stochastic disturbance term, the faster the city spreads out and thus the smaller is a reasonable maximum number of iterations.

Before examining the results obtained with this model, it will be useful to discuss briefly some features not included in it. First of all, as mentioned previously, there is no mechanism in the model described in this paper by which an occupied cell can return to a vacant state. In another version of the model, however, each cell, when it acquires an activity, also acquires a lifetime for that activity. With each iteration the lifetime is decremented automatically, but it is also augmented by an amount proportional to the potential calculated for the existing activity. Thus a cell well located for its current activity will be rejuvenated, while a poorly located cell will see its activity age. The probability of conversion to another activity, *ceteris paribus*, becomes greater as the remaining lifetime for the existing activity becomes shorter. If the lifetime falls to zero, the cell becomes vacant. This variant of the model permits the spatial reorganization of the city as the city grows - one common result is that some zones of the city in effect migrate. It also permits the urban cellular automaton to reach a state of self-organized criticality, if one is available, which the current model does not.

Secondly, all cells are assumed to be intrinsically identical - in other words, the city grows on the traditional isotropic plain. This is a useful characteristic when the primary goal is, as it is here, to study the general behaviour of the model and to search for general principles of urban development that may be thus revealed. But for applications to particular cities, it is necessary to take into account such factors as the physical characteristics of the land and legal restrictions on land use. A recently developed version of the model permits the integration of such factors into the calculation of the transition potentials. In effect, the model includes a simple GIS and calculates for each cell a suitability for each function. The inclusion of suitabilities in the calculation of transition potentials introduces the effects of physical and legal constraints on the development of the city, and incidentally thus reduces the effect of the random perturbation. This version of the model is currently being used to model land use on a Caribbean island, but has not yet been applied to an urban area.

Finally, there is no transportation network. Thus implicitly transport is of uniform quality throughout the region. This assumption is not too unreasonable for pre-industrial- or more precisely, pre-rail-cities, or for smaller modern cities with no rapid transit or express automobile routes. However, in general the transportation system is of major importance in determining urban form, and so it is clearly desirable to introduce one into the cellular automaton. A version of the model currently being developed includes a transportation network. Proximity to the network then becomes a factor in the calculation of the transition potentials. Thus the transportation network affects the pattern of urban growth.

10.3   SIMULATION RESULTS

Since the object of the present research is to explore general principles of urban spatial organization, the behaviour of the model under various combinations of parameter values was investigated. Thus, for example, the effects of different proportions of the three functions, of different growth rates, and of different levels of stochastic disturbance were examined. Most of these results were unexceptional and will not be described here. No calibration as such was performed since no particular urban area was modeled.

Nevertheless, some sets of parameter values give more reasonable results than others, while some yield patterns that bear no conceivable resemblance to any actual city. It was therefore encouraging that patterns of parameter values that represent known locational preferences were also the ones that generated the most realistic looking cities. Figure 1 shows five stages in the growth of one cellular city. For comparison, 1960 land use data for the city of Milwaukee, USA is shown on an 80 x 80 grid in Figure 2 (Passonneau and Wurman, 1966). Note, however, that four land use categories are used, the inclusion of an "other" category ensuring that the industry category on the Milwaukee map can not be the same as that for the cellular model.

The simulations discussed in this paper are based on a set of transition parameters that represent relatively conventional ideas about the locational behaviour of the various urban functions (White and Engelen, 1992). Specifically, it is assumed that commerce is strongly attracted to locate adjacent to existing commerce cells but is otherwise repelled by commerce in the area; and it is unaffected by industry, but is attracted to consumers, subject to a distance decay effect. Industry is unaffected by commerce, attracted to other industry in the immediate vicinity, and repelled by nearby housing but weakly attracted to housing located at a greater distance. Housing is repelled by adjacent commerce but attracted to it otherwise (subject to a distance decay effect), strongly repelled by nearby industry, and attracted to other housing, again with a distance decay effect.

The degree of random perturbation is an important control parameter. Increasing the value of n generally has the expected effect of increasing the complexity of the urban form. The relative importance of the stochastic effect is greatest during early iterations, when there are few occupied cells, so that the deterministic components of the transition potentials are all small and thus relatively similar in value, with the consequence that most variation in the final transition potentials is due to the stochastic term. As the number of occupied cells grows, the range of values calculated from the parameter matrix increases, thus reducing the relative importance of the stochastic term.

The particular form that a city evolves is extremely sensitive to even minor changes in parameter values, initial conditions, or change of seed in the random number generator. Thus, for example, in one pair of simulations, small changes in the parameter values representing the attraction of industry for itself shifted the major concentration of industry from the south to the north side of the city. And the two cities shown at iteration 40 in Figures 3 and 4 are identical in terms of both parameters and the series of random numbers generated for the stochastic disturbances. They differ only in their initial configurations, as shown. Nevertheless, in spite of this sensitivity, general properties of the spatial structure are robust. Two informal examples may be given. The largest concentration of commercial activity remains near the centre of the city, although it may relocate a short distance to follow a pronounced asymmetry in the location of housing, as can be seen in Figure 3. And industry almost always forms relatively compact clusters.

But there is a deeper sense in which the apparently highly diverse urban forms are similar. The urban forms are fractal structures, and, without major changes in parameter values, the fractal dimensions apparently vary only slightly from one city to another. The fractal nature of the patterns can be recognized in the scaling relationships that appear in both the size distribution of clusters of each activity and in the density gradient for each function. We will examine both of these relationships.

10.3.1   Cluster-Size Frequency Spectrum

Since fractal objects are self-similar at all scales - i.e. they are scale-free) - a log-log plot of the frequency of occurrence of clusters of each size class against cluster size will be linear if the object is a fractal. Otherwise a characteristic scale is present and the object is not fractal (Bak and Chen, 1989; Mandelbrot, 1983). In order to examine the size distribution of clusters, twelve simulations were run with identical parameter values, differing only in initial conditions (three configurations) and random generator seeds (four seeds). The frequency of occurrence of commercial clusters of each size class was accumulated over each set of four runs differing only in the random seed, and the frequencies were then plotted against size. One such plot is shown in Figure 5a. In each case, the relationship was linear. Only commercial clusters were examined since there were too few clusters of the other functions. Clusters were defined using only horizontal and vertical adjacencies for the sake of consistency with the procedure of other authors.

The clusters measured are not necessarily stable, since in this model cells in other states can always be converted to commercial; thus the relationship described is that existing at a particular time: iteration 40. Nevertheless, the slopes, approximately 1.71 in the case illustrated, remain nearly constant as the iterations proceed, so the relationship is essentially unchanged as the city grows. For comparison, a similar plot for the commercial clusters appearing in the map of Milwaukee (Figure 2) is shown in Figure 5b. The relationship there is also linear, with a slope of 1.75. Most other cities examined also show a linear size-frequency relationship, although actual slopes differ from city to city (White and Engelen, 1992). Thus by this measure, real cities as well as those produced by the cellular model have a fractal structure.

10.3.2   Area-Radius Scaling

The cellular cities described here may be thought of as what Mandelbrot (1983) calls a "dust", that is a scattering of points. A fractal dust has the property that if the mass of the object, that is, the number of points, N, is measured within a radius r of a given point, then as r is increased, the mass, N, increases as r^D where D is the fractal dimension. The log-log plot of such a relationship is linear, with slope D. For $D < 2$ the object is increasingly sparse at larger distances.

The cumulative values of N by two-unit increments in r were established for one set of the twelve simulations just described, runs differing only in the random seed used. Mean values of N for commerce for the four cities are plotted against r in Figure 6a. The relationship is highly linear out to a radius of 16, with a slope for the trend line for this portion of 1.09. Beyond $r=16$, the relationship is also approximately linear, but with a much flatter slope of 0.30. The points beyond $r=26$ are not shown, since values for those zones are dominated by edge effects. The implication is that commerce cells are distributed fractally, but that two fractals, with different dimensionality, are involved.

Plots were also made for industry and housing. These were also in general highly linear but kinked, but the kinks generally did not occur at the same radii. The plots for industry and housing for the same set of cities shown in Figure 6a are shown in Figures 6b and 6c respectively. With respect to the inner portion of the kinked plots, in every case the slopes for commerce are less than those for industry, which are in turn less than those for housing. For the outer portion, the slopes for housing are less than those for industry.

The implication is that these cellular cities have a tendency to concentric land use zones, with commerce concentrated in the centre. Analogous plots for Milwaukee are shown in Figure 7. The kink is present in the industry plot, but absent in commerce and housing. It is likely that the kink point for these activities occurs beyond the edge of the map, since the map does not include the peripheral portions of the city.

Finally, the urbanized area itself seems to be a bi-fractal structure. All cities generated by the cellular model show an inner zone with a dimensionality of approximately 1.9 as measured by the slope of the area-radius relationship, and an outer zone with a dimensionality typically within the range of 1.1 to 1.3. Frankhauser (1991) and Frankhauser and Sadler (1991) have examined the urbanized areas of a number of cities on several continents and found that in virtually every case the cities exhibited a bi-fractal structure with dimensionalities close to these values. The kink point apparently represents the edge of the area in which the urbanization process is essentially complete, with the outer zone being the area in which the urbanization is still in progress. Note that in the inner zone, although the urbanization process is essentially complete, not all of the land is occupied; the city retains the form of a Sierpinski carpet.

10.4   DISCUSSION

In fractal cities detail is present at all scales. Intuitively, this is important. Comprehensively designed urban areas which lack detail at some scales are typically spoken of as sterile - people perceive something to be missing. People do not live in the housing stock in general, as it would be represented by a state variable in a dynamic model; they live in particular houses or apartments, with particular surroundings, in neighbourhoods with particular characteristics. For individuals, with their individual needs, it is important to find sufficient variety in these particularities, because then they will be able to locate situations that suit them. Jacobs (1961), an acute observer of the way cities really work, described in great detail the need for structural heterogeneity in a city. The situation is closely analogous to that of patchiness in natural environments. The average quality of habitat in an area might not be suitable for a particular species, but if individuals are able to locate and remain in patches of above-average quality, then the species can survive there. Many more things are possible in a complex environment than in a simple one.

But more formally, what is the significance of the fact that cities are apparently multi-fractal objects? Recent work in the theory of cellular automata and other connectionist models throws some light on the nature of the structures that evolve in a city insofar as the cellular automata models capture something of the real process. Kauffman (1990), analyzing the properties of arbitrary Boolean (binary) networks, systems that have much in common with cellular automata, and that can also be considered as models of disordered dynamical systems, finds that under certain circumstances stable structures develop. Specifically, certain kinds of transition rules permit groups of cells jointly to become locked in a particular state, so that they are, in effect, isolated from the behaviour of the rest of the system. In the absence of these structures, the behaviour of the network appears essentially chaotic, with an extreme sensitivity to small perturbations. But when the ensemble of transition rules applied in the network includes a sufficiently high proportion of rules which permit cells to become locked in a particular state - and the more connected the system, the higher the proportion required - then islands of stability link up into a reticulation which percolates (that is, permeates through the entire system) and provides a skeleton of stability. The previously chaotic behaviour is then confined to isolated areas which cannot affect each other, and consequently becomes simplified to relatively short limit cycles. In short, systems in which frozen elements percolate are highly structured, with large basins of attraction around a relatively small number of attractors.

A percolating fixed structure furthermore seems to be necessary for the structural evolution of the automaton, at least if such evolution is to be achieved by hill-climbing on a fitness landscape. In the absence of such a structure, the fitness surface is uncorrelated - it has a rough and random topography. In other words, similar systems do not tend to have similar fitnesses, and a given structure carries no information about the fitness of even very similar structures, so evolution by selection for fitness is not possible. Systems with percolating frozen structures, however, tend to have highly correlated fitness landscapes, and so may evolve. As the frozen structure becomes more extensive than its minimum percolation size, however, the system looses evolutionary flexibility.

Langton (1990), in related work with cellular automata, has shown that persistent complex or self-organizing behaviour essentially occurs only at a phase transition between stable behaviour and a chaotic regime. The type of behaviour exhibited by an automaton depends on the proportion of possible states for the neighbourhood of a cell which will result in that cell being set to the quiescent state. If that proportion is high, then the automaton will quickly move to a point or limit cycle attractor. If the proportion is lower than a critical value, then the automaton exhibits chaotic behaviour, so that it has an apparently random structure. But at the critical point between the stable and chaotic regimes, transients of arbitrary length and complexity develop, so the system exhibits very complex - but not random - structure. If the structure is characterized by its Shannon entropy, H, these maximally complex structures are found to correspond to an intermediate value of H, with complexity peaking sharply at the particular value of H marking the phase transition between the ordered and chaotic states.

At the critical point, structure is present at all scales, and is correlated at arbitrarily large distances, so that there is global as well as local organization. Furthermore, in this critical area, the complexity of the structure is exponentially related to system size, whereas in both the stable and chaotic regimes, behaviour is largely independent of system size. The conclusion here, then, is essentially the same as that reached by Kauffman: the appearance of complex structure by self-organization depends on a fine balance between order and disorder. Evolved structure may, in fact be characterized as an elaborate combination of order and disorder.

Returning to the cellular cities described above, it is evident that much of the discussion of Kauffman and Langton applies to them. Cities with a regular geometric form would have low values of the Shannon entropy, H, and those with a largely chaotic or random structure would have a high value. Cellular cities, and real cities as well, however, are characterized by intermediate values of H, because they have a very complex form, with structure evident at various scales, as measured by the various fractal dimensions. And fractal dimensions or power-law spectra of features seem to be characteristic of systems just at the boundary between order and chaos (Kauffman, 1990;

Bak and Chen, 1989; Bak, Chen, and Creutz, 1989). Furthermore, the fact that the fractal dimensions for different land uses are systematically different reveals a tendency toward concentric zones of land use underlying the cluster pattern which is most evident on the maps. There is thus a global correlation in spite of the local nature of the rules generating the patterns - another characteristic, according to Langton (1990), of the phase transition point.

Cellular cities do not start out with a complex structure, however. They evolve it. In the initial iterations, the behaviour of the automaton is almost entirely determined by the initial conditions and the random fluctuations. The structure is essentially random, characterized by a high Shannon entropy. A small change in initial conditions, or random perturbations, will, as remarked above, lead to a very different configuration. However, the system soon organizes itself so that larger scale structures develop, representing such features as industrial districts and commercial centres; and from this point on, as the city grows and changes, and new structures appear at ever larger scales, the general nature of the form, as measured by the fractal dimensions, is maintained. Only on the fringe, where the urbanization process is nascent, does the structure remain largely random, and highly sensitive to perturbations.

10.5   CONCLUSION

The model discussed in this paper is a very simple one. Nevertheless it produces relatively realistic and surprisingly complex representations of urban areas. The fractal structure of the cities generated is evident both in the non-spatial feature of the size spectrum of commercial clusters, and in the spatial features of the area-radius relationships and the urban boundary. Indeed, the structure should be described as multi-fractal, since each of the three land use types has a characteristic pair of fractal dimensions, with an abrupt transition from one dimension to another at a particular radius. Data from a sample of U.S. cities displays the same multi-fractal structures. Results from research by other workers in the field of cellular automata and Boolean networks suggests that cities, as evolved structures, are necessarily and inherently complex fractal objects.

ACKNOWLEDGEMENTS

We thank Serge Wargnies and Inge Uljee for their help with this research. This work was supported by the Research Institute for Knowledge Systems, Maastricht, the Social Sciences and Humanities Research Council of Canada, and Memorial University of Newfoundland.

REFERENCES

Allen, P., 1983, Self-Organization and Evolution in Human Systems, (Crosby, R. ed.), **Cities and Regions as Nonlinear Decision Systems**, Westview Press, Boulder, pp. 29-62.

Allen, P., G. Engelen, and M. Sanglier, 1984, Self-Organizing Dynamic Models of Human Systems, (Ferhland, E. ed.), Macroscopic to Microscopic Order, **Synergetics**, vol. 22), Springer, pp. 150-171, Berlin.

Allen, P. and M. Sanglier, 1979, A Dynamical Model of Growth in a Central Place System, **Geographical Analysis**, vol. 11, pp. 256-272.

Bak, Per and Chen, Kan, 1989, The Physics of Fractals, **Physica D**, vol. 38, pp. 5-12.

Bak, Per; Chen, Kan; and M. Creutz, 1989, Self-Organized Criticality in the 'Game of Life. **Nature**, vol. 342, pp. 780-782.

Batty, M., 1991a, Cities as Fractals: Simulating Growth and Form, (A.J. Crilly, et. al. eds.), **Fractals and Chaos**, Springer-Verlag, pp. 43-69.

Batty, M., 1991b, Generating Urban Forms from Diffusive Growth, **Environment and Planning A**, vol. 23, pp.511-544.

Batty, M., and P. Longley, 1987, Fractal-Based Description of Urban Form, **Environment and Planning B**, vol. 14, pp. 123-134.

Batty, M., P. Longley, and S. Fotheringham, 1989, Urban Growth and Form: Scaling, Fractal Geometry, and Diffusion-Limited Aggregation, **Environment and Planning A**, vol. 21, pp. 1447-1472.

Couclelis, H., 1985, Cellular Worlds: A Framework for Modeling Micro-Macro Dynamics, **Environment and Planning A**, vol. 17, pp. 585-596.

Couclelis, H., 1988, Of Mice and Men: What Rodent Populations Can teach Us About Complex Spatial Dynamics, **Environment and Planning A**, vol. 20, pp. 99-109.

Couclelis, H., 1989, Macrostructure and Microbehavior in a Metropolitan Area, **Environment and Planning B**, vol. 16, pp. 141-154.

Engelen, G., 1988, The Theory of Self-organization and Modelling Complex Urban Systems, **European Journal of Operational Research**, vol. 37, pp.42-57.

Fotheringham, S., M. Batty, and P. Longley, 1989, Diffusion-Limited Aggregation and the Fractal Nature of Urban Growth, **Papers of the Regional Science Association**, vol. 67, pp. 55-69.

Frankhauser, P., 1991, Aspects fractals des structures urbaines, **L'Espace Geographique**, pp. 45-69.

Frankhauser, P. and R. Sadler, 1991, Fractal Analysis of Agglomerations, **Proceedings of the Second International Colloquium of the Sonderforschungsbereich 230: Naturliche Konstruktionen.**

Grassberger, P., 1991, La Percolation ou la Geometrie de la Contagion, **La Recherche**, vol. 22, pp. 640-646.

Hillier, W. and J. Hanson, 1984, **The Social Logic of Space**, Cambridge University Press.

Jacobs, J., 1961, **The Death and Life of Great American Cities**, Random House, New York.

Kauffman, S.A., 1990, Requirements for Evolvability in Complex Systems: Orderly Dynamics and Frozen Components, **Physica D**, vol. 42, pp. 135-152.

Langton, C., 1990, Computation at the Edge of Chaos: Phase Transitions and Emergent Computation, **Physica D**, vol. 42, pp. 12-37.

Mandelbrot, B., 1983, **The Fractal Geometry of Nature**, W. H. Freeman and Co., New York.

Markus, M. and B. Hess, 1990, Isotropic Cellular Automaton for Modelling Excitable Media, **Nature**, vol. 347, pp. 56-58.

Passonneau, J. and R. Wurman, 1966, **Urban Atlas: 20 American Cities**, MIT Press.

Phipps, M., 1989, Dynamical Behaviour of Cellular Automata Under Constraint of Neighbourhood Coherence, **Geographical Analysis**, vol. 21, pp. 197-215.

Pumain, D., T. Saint-Julien, and L. Sanders, 1987, Applications of a Dynamic Urban Model, **Geographical Analysis**, vol. 19.

Tobler, W., 1979, Cellular Geography, (S. Gale and G. Olsson, eds.), **Philosophy in Geography**, pp. 379-386.

White, R., 1977, Dynamic Central Place Theory: Results of a Simulation Approach, **Geographical Analysis**, vol. 9, pp. 227-243.

White, R., 1978, The Simulation of Central Place Dynamics: Two Sector Systems and the Rank-Size Distribution, **Geographical Analysis**, vol. 10, pp. 201-208.

White, R., 1984, Principles of Simulation in Human Geography, (G. Gaile and C. Wilmott, eds.), **Spatial Statistics and Models**, D. Reidel, Dordrecht, pp. 384-416.

White, R. and G. Engelen, 1992, Cellular Automata and Fractal Urban Form: A Cellular Modelling Approach to the Evolution of Urban Land Use Patterns, **Environment and Planning A**, in press.

Wilson, A., 1978, Spatial Interaction and Settlement Structure: Toward an Explicit Central Place Theory, (A. Karqvist, et. al., eds), **Spatial Interaction, Theory, and Planning Models**, North Holland, Amsterdam, pp. 137-156.

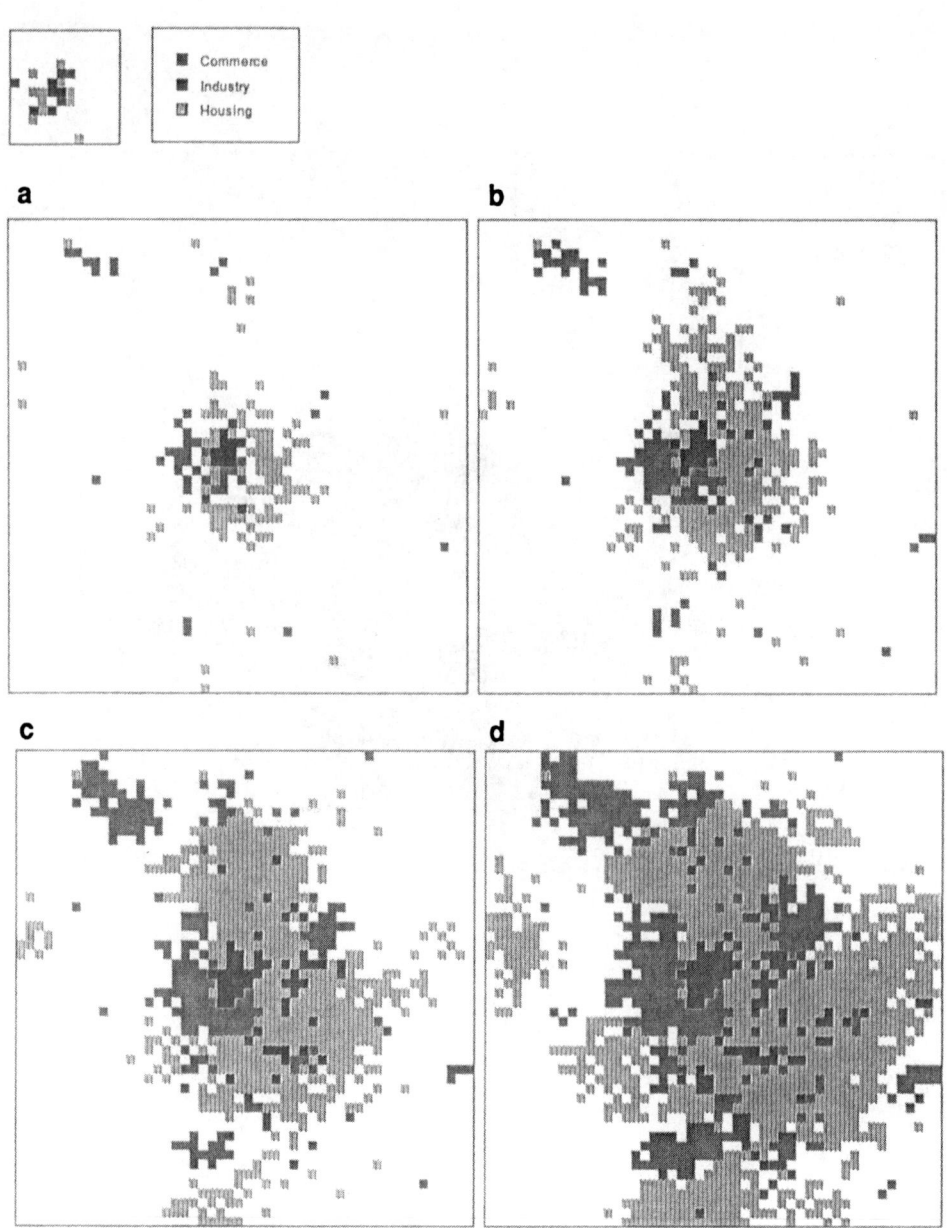

Figure 1　**Stages in the Growth of a Cellular City.** Iterations 10, 20, 30, 40 with initial configuration at upper left.

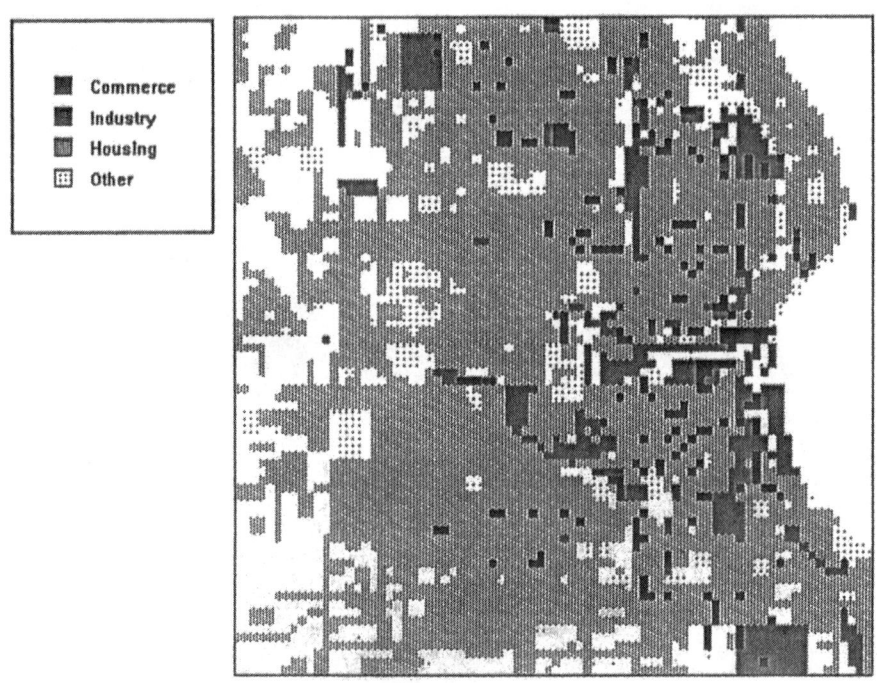

Figure 2 Land Use in Milwaukee, USA, 1960.

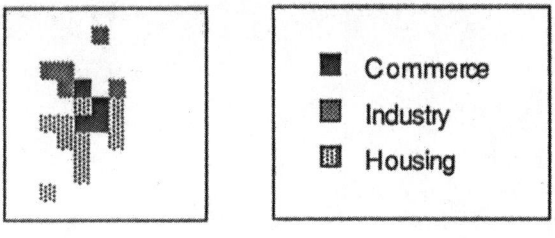

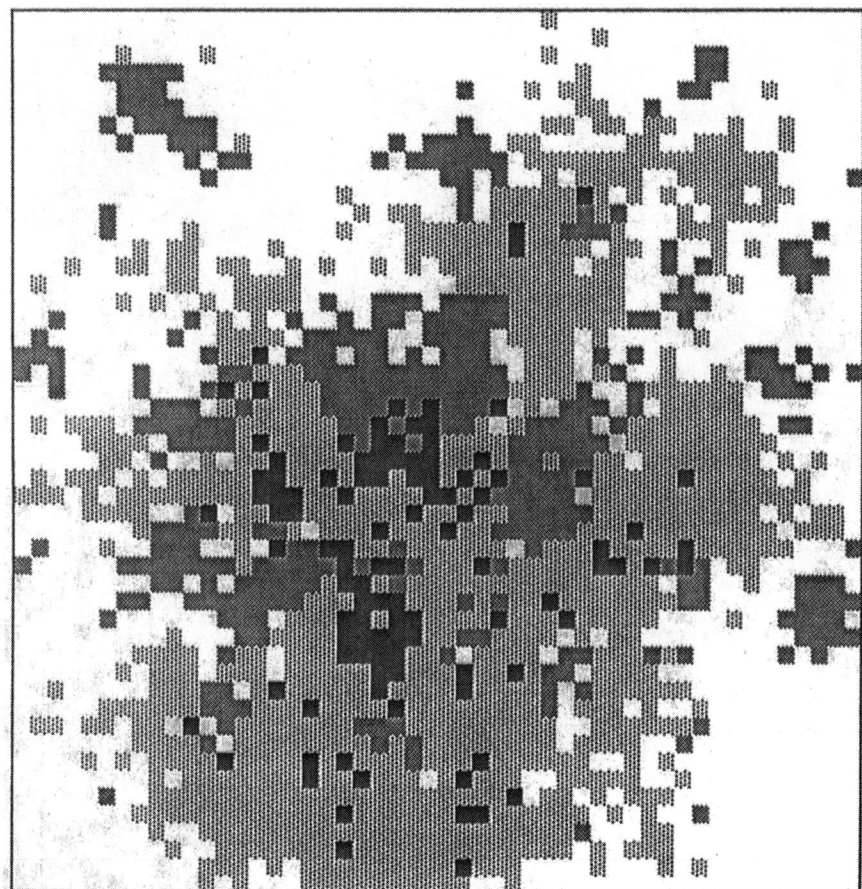

Figure 3 **A Cellular City at Iteration 40.** The Commercial Centre has shifted south.

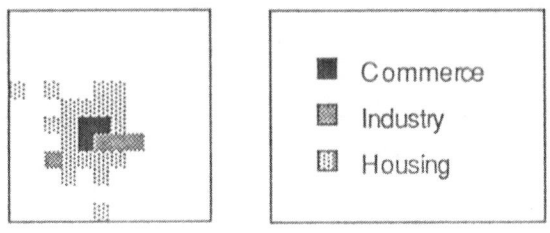

Figure 4 **Alternate Cellular City at Iteration 40.** This city eas generated by the same parameters as those used for city in Figure 3; only the initial configuration differed.

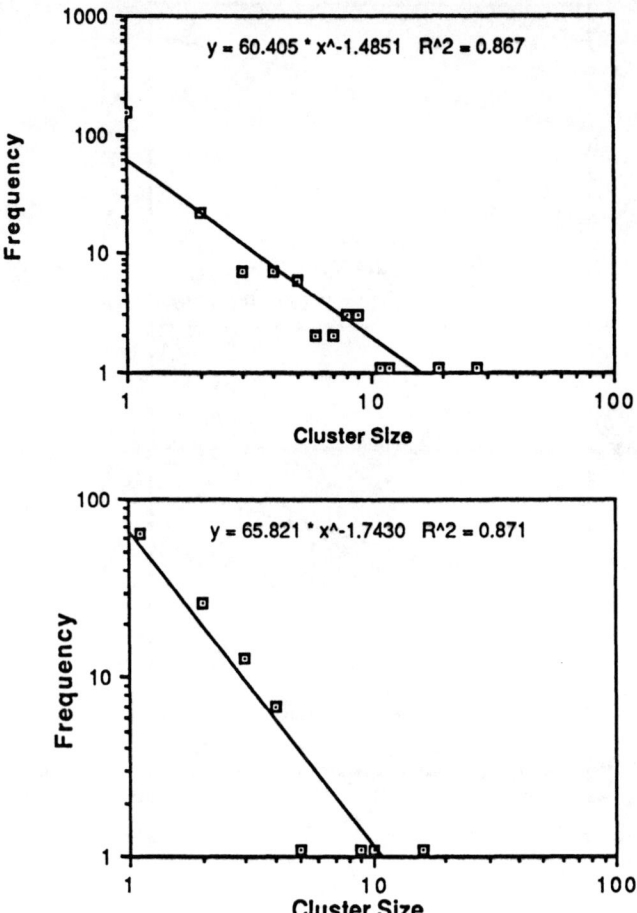

Figure 5 **CLUSTER SIZE FREQUENCY SCALING:** Aggregate data from four simulations differing only in stochastic perturbation.
a. **(top)** **Cellular cities.**
b. **(bottom)** **Milwaukee, 1960.**
Source of data: Passonneau and Wurman, 1966)

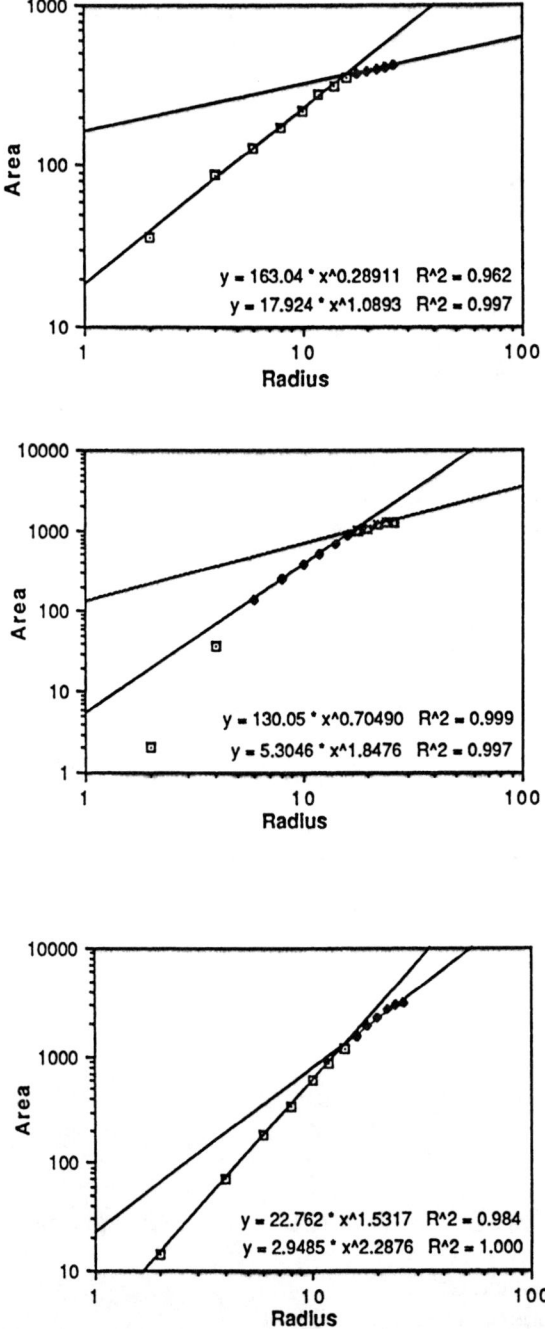

Figure 6 AREA-RADIUS SCALING FOR CELLULAR CITIES:
Aggregate data from four simulations differing only in stochastic perturbation.
a. (top)   Commerce
b. (middle) Industry
c. (bottom) Housing

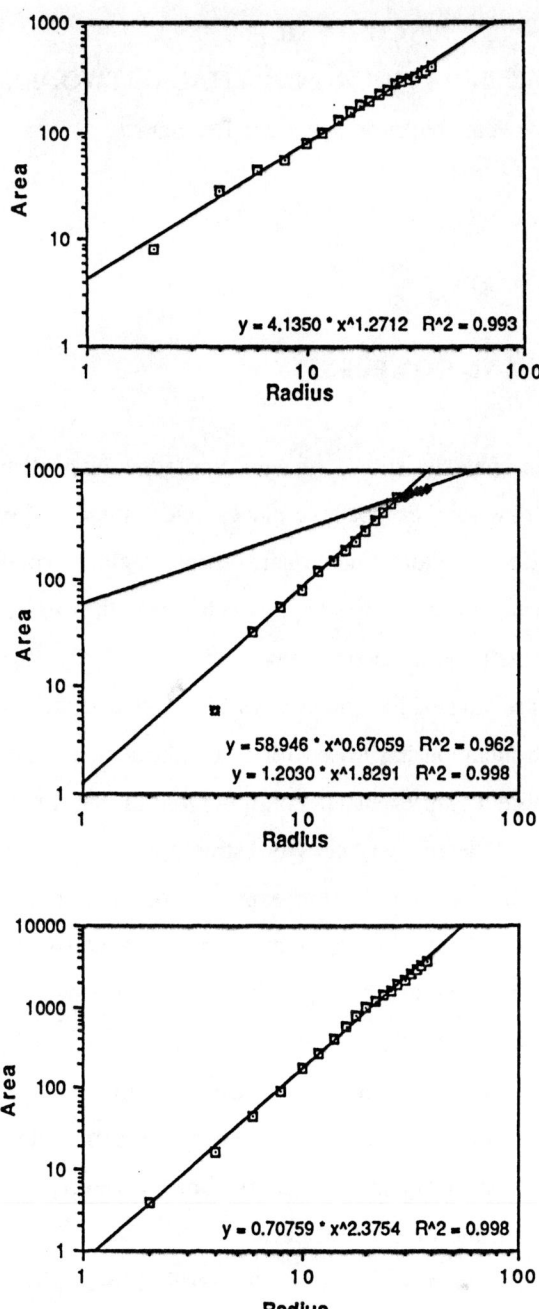

Figure 7 **AREA-RADIUS SCALING FOR MILWAUKEE.**
a. (top) Commerce
b. (middle) Industry
c. (bottom) Housing
Source of data: Passonneau and Wurman, 1966.

CHAPTER 11
COMPLEX BEHAVIOUR IN SPATIAL NETWORKS
Peter Nijkamp and Aura Reggiani

11.1 NETWORKS AND COMPLEXITY

Connectivity is a basic characteristic of societies and economies all over the world. Communications and transactions cannot take place, unless there is (social or spatial) connectivity between different actors (individuals, firms, regions, countries etc). For example, social contacts presuppose a language which forms the vehicle for verbal or written interactions. International or interregional trade takes for granted the existence of a transportation infrastructure which connects various countries or regions.

In the past years much attention has arisen for a specific kind of connectivity which manifests itself as a more comprehensive configuration in the form of **networks**. Networks represent an ordered connectivity structure for communication and transportation which is characterized by the existence of main nodes which act as receivers or senders (pull and push centers) and which are connected by means of corridors or edges. If the nodes are having a hierarchical structure, also the edges reflect a similarly ordered organisation, as is for instance witnessed by the 'hub and spokes' system in modern airline systems and coastal transport.

Clearly, the **configuration and order** in a network deserve thorough scientific analysis, e.g. by using graph theory or Boolean algebra. However, a network derives its existence and merit from the functions it performs. In other words, in investigating the significance of networks we have to look at the services they provide (in terms of internodal flows and their contribution to systems's performance). Such services are not static, but are determined by behavioural mechanisms (e.g. individual or group decisions), policy choices (e.g., on network design or operation) or on bottlenecks (e.g., capacity constraints). The resulting flows may have stable patterns, but may also suddenly become unstable (e.g., in case of sudden bifurcations). Thus the outcomes of network operations may be either stable or unstable (e.g., in case of chaotic evolution).

Networks are usually regarded as **complex** systems; the adjective 'complex' is often used in network analysis. From a philosophical perspective complexity involves notions of non-intuitive system's behaviour, in particular evolutionary patterns of connections between subsystems for which prediction of system's behaviour is hard without substantial analysis or computation, especially of decision-making structures that make the effects of individual choices difficult to forecast (Casti, 1979, p.97). Following Casti (1979), we will define here different types of complexity, which clearly can influence the evolution of a network:

a)   **Static complexity**

This phenomenon refers mainly to the structure of the system while neglecting any dynamical or computational considerations. It concerns in particular a configuration where the components of a system are put together in an intricate and interrelated way. In this framework several types of complexity may arise, which may be derived from the following forms:
- **hierarchical** structure: the number of hierarchical levels can roughly measure the static complexity of a system or network
- **connective** pattern: the manner in which the components or subsystems of a process are connected - e.g., by means of q-analysis - may represent a complexity measure
- **variety** of components: variety emerges when a system is able to manifest itself in different behavioural modes or outputs
- **strength of interactions**: this phenomenon is strictly connected with the hierarchical structure and connective pattern, since it offers a measure of the 'practical' complexity.

b)   Dynamic complexity

This phenomenon refers to the dynamic behaviour of a low static system and may originate from both a low and high static complexity. It is essentially linked with the ability to predict system behaviour which can roughly be measured by:

- **computational** complexity: this refers to the question how many steps a given algorithm requires for computing a particular function
- **evolutionary** complexity: this phenomenon describes the ability of a system to interact with its environment as well as to adjust to it. In this framework, chaos models may be regarded to have a high level of systemic complexity, since their ability to mirror the behavioural (random) mechanisms of a certain phenomenon is extremely high.

In a recent book (see Nijkamp and Reggiani, 1992a) the issue of spatial interaction, evolution and chaos has extensively been dealt with. Various simple illustrations were used to underscore the importance of chaos in nonlinear dynamic spatial systems. In the present chapter we will elaborate on these issues by paying attention to the dynamic (stable or unstable) evolution of comprehensive spatial network systems. The stability conditions will also be investigated, while next the degree of spatio-temporal autocorrelation in the state variables will be examined in greater detail, with a particular view on their stability over time in case of chaotic spatial evolution. The analysis will be based on a simplified multiregional network model.

11.2   NETWORKS AS SPACE - TIME SYSTEMS

Spatial connectivity can be characterized by two features (see Bennett and Chorley, 1978):

- **specificity**: phenomena and systems have a distinct locational component.
- **relativity**: phenomena and systems have space-time relationships, so that external interactions impact upon their states.

Formally, the way in which state variables characterizing phenomena and systems are related to one another can be described by means of operators linking inputs to outputs. Such operators are often called **transfer** functions.

Space-time evolution can be represented by assuming the following transfer function for a vector set of endogenous variables $\underline{x}$ and exogenous variables $\underline{z}$:

$$\dot{\underline{x}}_r = f(\underline{x}_r, \underline{z}_r ; \underline{x}_{r'}, \underline{z}_{r'}) \tag{11.1}$$

where the subscript r refers to the region r concerned, and r' to all other relevant regions. It is clear that the transfer function (11.1) leads to stable evolution (or at least regular evolution), if f is linear. In case of nonlinearities, various types of irregular (non-cyclical or chaotic) patterns may emerge. The conditions under which stable or unstable evolution will develop deserve more attention and will be discussed later on.

Another question concerns the type of spatial connectivity pattern which impacts on the resulting spatial interaction flows. Bennett and Chorley (1978) make a useful distinction into four space-time configurations:
- **barriers**: the interaction between places is prevented or discouraged in some way
- **hierarchies**: spatial interaction is controlled by a hierarchical dependence
- **networks**: spatial interactions are determined by existing physical or non-physical (e.g., communication) infrastructure pathways
- **contiguities**: spatial interactions are determined by linkages between adjacent places.

It should be added that in reality various mixed configurations may occur. For instance, network impacts may generally be supposed to be more intensive for adjacent places on a network than for distant places. This means that the idea of **spatial contiguity analysis** can also be applied to points in a network (see Paelinck and Nijkamp, 1975). First-order contiguity would then occur for all points in a spatial network which have a first-order Boolean connectivity. Second-order connectivity then emerges for a second-order Boolean representation etc.

We will adopt in our approach the Moran coefficient (Moran, 1950) as defined in

Cliff and Ord (1975, 1981):

$$M = n \sum_i w_{ij} e_i e_j / W \sum_i e_i^2 \qquad (11.2)$$

where:

w_{ij} is the weight (strength of the link) between centre i and centre j

$e_i = x_i - \bar{x}$

$e_j = x_j - \bar{x}$

n is the number of centres

$$W = \sum_i \sum_j w_{ij} \quad (i \neq j) \qquad (11.3)$$

It should be noted that a slightly different formulation of the Moran coëfficient is the Geary coëfficient (Geary, 1954):

$$G = [(n-1)/2W] \sum_i \sum_j w_{ij} (x_i - x_j)^2 / e_i^2 \qquad (11.4)$$

Equations (11.2) and (11.4) are interesting since it expresses the possibility of different formulations for the weight. In particular, two frequently used weighting schemes are:

$w_{ij} = 1$, if centres i and j share a common edge

$w_{ij} = 0$, otherwise $\qquad (11.5)$

$w_{ij} = 1/d_{ij}$, where d_{ij} is the distance between the centroids of areas i and j. $\qquad (11.6)$

In particular both Moran and Geary considered only binary weights as expressed in (11.5), while Cliff and Ord (1981) extended their results to generalized weights.

Further extensions of spatial contiguity analysis towards **spatio-temporal autocorrelation and space-time lag operators** can be found amongst others in Anselin (1988), Griffith (1992) and Hordijk and Nijkamp (1977). The interesting point in the above approach is that it allows for the calculation of average spatial influences of a given state variable in a dynamic spatial interaction system. This might then be computed over various time periods in the simulation process of a dynamic (linear or nonlinear) model. The time pattern of such a set of spatial autocorrelation coefficients might then be represented by means of a time correlogram (see Figure 1).

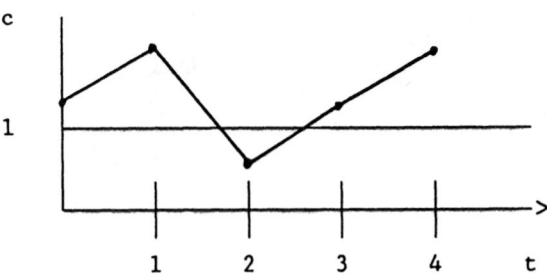

Figure 1   A time correlogram

11.3   A SIMPLE SPATIAL NETWORK MODEL

In our exploratory analysis we will assume a simplified spatial network configuration which for the time being is supposed to be composed of three nodes in a fully interwoven network (see Figure 2). Later on we will also deal with higher-order network configurations.

It is assumed however, that not all nodes are equally important nor that their mutual influences are the same. Some may have a negative influence upon others, and others a reverse influence. To some extent such a system can be described as a predator-prey system (see Nijkamp and Reggiani, 1992b).

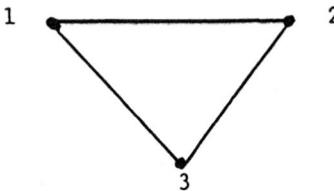

Figure 2   An interwoven three node network

We assume now the following main features for this network system, on the basis of a diffusion model with spatial and temporal elements which originates from previous work by Paelinck and Nijkamp (1975):
- explicit introduction of **time elements**
- explicit introduction of a **spatial interwovenness** between specific nodes (e.g., by means of spin-off and spill-over effects)
- **functional relationships** between all relevant phenomena governing a **diffusion process** in a network.

Then a simplified model with the above features may be composed from the following set of three structural dynamic equations related to three nodes n (n,n'=1,2,3), as illustrated in Figure 2:

$$A^n_{t+1} = \lambda^n A^n_t (1 - A^n_t) \qquad (11.7)$$

This first equation describes the **attractiveness** A of node n. It assumes that the development of A follows a (standard) logistic evolution (in relative terms). It seems plausible to take for granted that the dynamic evolution of nodal attractiveness cannot exceed a critical threshold level (i.e.,1), although it may assume a wide spectrum of dynamic behaviour (including feedbacks) (see also Batten, 1987). This particular shape of the equations gives also the possibility of unstable and chaotic movements depending on the values of λ^n (see May, 1976).

The second equation starts from the definitional equation

$$K^n_{t+1} \equiv K^n_t + I^n_{t+1} \qquad (11.8)$$

where K^n_t represents the **capital stock** in centre n at time t. I^n_{t+1} is the **investment** function in node n at time t+1 which can be represented as follows (see Paelinck and Nijkamp, 1975):

$$I^n_{t+1} = \alpha^n_t A^n_t + \sum_{\substack{n'=1 \\ n' \neq 1}}^{N} \beta^{nn'}_t K^{n'} + \gamma^n_t Y^n_t \qquad n = 1,2,3 \qquad (11.9)$$

In (11.9) we assume that the investment decision in node n, at time t+1, is a linear function (i) of some attractiveness of node n (denoted by $\alpha^n_t A^n_t$ where α^n_t is a parameter), (ii) of an aggregated spill-over effect from capital stocks in the remaining centres in period t (denoted by the second term in (11.9) (where $\beta^{nn'}_t$ is the spatial spill-over effect in the network), and (iii) of production Y in node n at time t (which is regulated by a growth parameter γ^n_t).

Now equations (11.8) and (11.9) can be integrated in a single equation as follows:

$$K^n_{t+\overline{1}} = \alpha^n_t A^n_t + \sum_{\substack{n'=1 \\ n' \neq 1}}^{N} \beta^{nn'}_t K^{n'}_t + \gamma^n_t K^n_t \qquad n = 1,2,3 \qquad (11.10)$$

which represents now the second 'evolutionary' equation of node n (besides equation (11.7)).

The final equation describes the evolution of **production** Y in node n:

$$Y^n_{t+1} = \sum_{n'=1}^{N} \delta^{nn'}_t P^{n'}_t + \sum_{n'=1}^{N} \epsilon^{nn'}_t Y^{n'}_t + b^n_t \qquad n = 1,2,3 \qquad (11.11)$$

This last equation assumes that the income of node n is a function of the volume of population in all contiguous centres in the previous period t as well as of the volume of production in all centres in the same period t. For the moment we assume that this relationship is linear. The parameters $\delta^{nn'}_t$ and $\epsilon^{nn'}_t$ are again 'space-time' coefficients taking into account the spatial contiguity effects described in Section 11.2. The last term b_t^n in (11.11) represents here a 'control' parameter which can influence the evolution of production (e.g., capacity in a transport system, profit tax, etc.). Thus equations (11.7), (11.10) and (11.11) do now represent the 'space-time' evolution of the three variables A_t^n, K_t^n and Y_t^n related to three nodes n.

It should be noted that, depending on the assumption whether n' is an **attraction centre** (with pull forces negatively affecting surrounding areas) or a **growth centre** (with push forces positively affecting adjacent regions), the 'space-time' coefficients $\beta_t^{nn'}$, $\delta_t^{nn'}$ and $\epsilon_t^{nn'}$ are negative or positive, respectively.

11.4 STABILITY AND COMPLEXITY IN A SIMPLE 3-D HIERARCHICAL SYSTEM

After the preliminary notions on systemic complexity discussed in Section 11.1 we will show - on the basis of the example developed in Section 11.3 - how different aspects of complexity can affect the dynamic behaviour of a network. For this purpose some **simulation** experiments will be carried out. In particular we will deal with the relationship stability/complexity by considering various relevant aspects of system complexity described above. In particular, we will consider respectively different kinds of hierarchy for a 3-D (three dimensional) fully connected spatial network having three nodes (see Figure 2):

a) centre 1 is an attraction centre (and centres 2 and 3 are growth centres);
b) centres 1 and 2 are attraction (and hence competing) centres;
c) centres 1, 2 and 3 are three attraction (and thus competing) centres.

We will in particular take into account the possibility for variable A (nodal attractiveness) of generating chaotic values.

a) Centre 1 as attraction centre

In this simulation experiment we will assume the following parameter values in the equations (11.7), (11.10) and (11.11):

$\lambda^1 = \lambda^2 = \lambda^3 = 3.9$

$\alpha^1 = 0.01$	$\alpha^2 = 0.03$	$\alpha^3 = 0.07$
$\beta^{12} = 0.05$	$\beta^{13} = 0.05$	$\gamma^1 = 0.01$
$\beta^{21} = -0.03$	$\beta^{23} = 0.04$	$\gamma^2 = 0.03$
$\beta^{31} = -0.03$	$\beta^{32} = 0.002$	$\gamma^3 = 1.1$
$\delta^{11} = 0.004$	$\delta^{12} = 0.005$	$\delta^{13} = 0.005$
$\delta^{21} = -0.017$	$\delta^{22} = 0.006$	$\delta^{23} = 0.01$
$\delta^{31} = -0.04$	$\delta^{32} = 0.004$	$\delta^{33} = 0.02$
$\epsilon^{11} = 0.05$	$\epsilon^{12} = 0.01$	$\epsilon^{13} = 0.01$
$\epsilon^{21} = -0.04$	$\epsilon^{22} = 0.05$	$\epsilon^{23} = 0.01$
$\epsilon^{31} = -0.02$	$\epsilon^{32} = 0.03$	$\epsilon^{33} = 0.05$
$b^1 = 1$	$b^2 = 0.5$	$b^3 = 0.009$

with the initial values for the variables:

$A^1 = A^2 = A^3 = 0.1$

$K^1 = 0.16 \quad K^2 = 0.04 \quad K^3 = 0.1$

$Y^1 = 0.05 \quad Y^2 = 0.02 \quad Y^3 = 0.1$

and

$P^1 = 0.1 \quad P^2 = 0.25 \quad P^3 = 0.2$

It should be noted that the original values of the variables A^n, K^n, Y^n, P^n, have been multiplied here with a factor equal to 0.001 in order to get values compatible with the values of $A^n < 1$, as expressed by the logistic equation (11.7).

The numerical simulations show a fluctuating pattern for the variables K^1, K^2, K^3, (see Figure 3). We do not display here the evolution of the variables Y^1, Y^2, Y^3, since their pattern is clearly stable (as we can see from the linear equation (11.11)).

It is interesting to observe that, if we consider a non-chaotic value for the variable A^n (e.g., $A^1 = A^2 = A^3 = 1.1$), fluctuations for capital still emerge (see Figure 4).

b) **Centres 1 and 2 as attraction centres**

In this simulation we will use the same parameter values as adopted in Section

11.4a, except negative signs for the diffusion coefficients β^{12}, β^{32}, δ^{12}, δ^{32}, ϵ^{12}, ϵ^{32}.

The emerging pattern for the variable K is clearly stable despite the chaotic regime of the variable A^n (see Figure 5). It seems surprising that, by increasing - in this simple 3-D network - the number of competing nodes, the fluctuations disappear. However, this result seems to reinforce the Smith (1974) hypothesis that in an ecological web the increase of competing species leads to more stability.

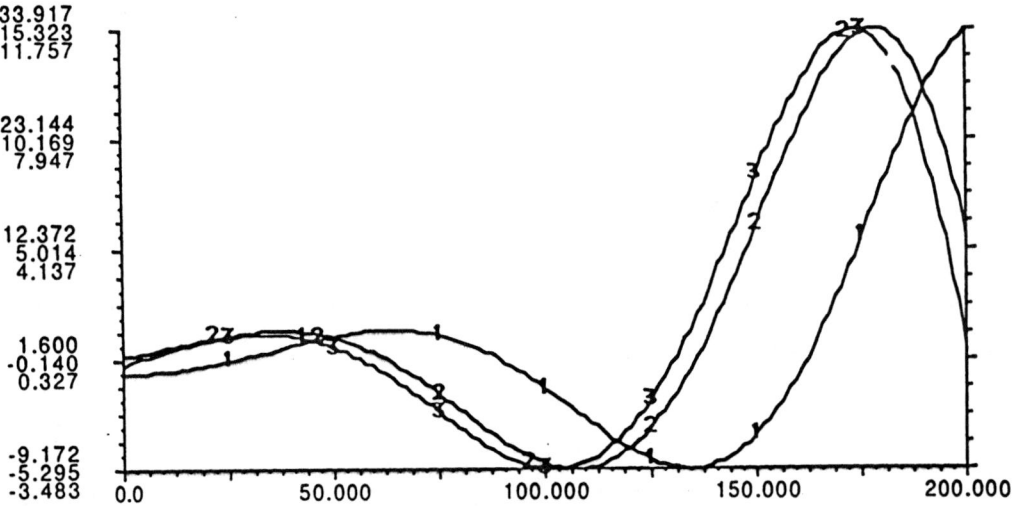

Figure 3 Fluctuations of capital K in a 3-D spatial network with one node as attraction centre and two nodes as growth centres (under a chaotic regime for the attractiveness); y-axis = K^1, K^2, K^3 (in descending order from the top), x-axis = time

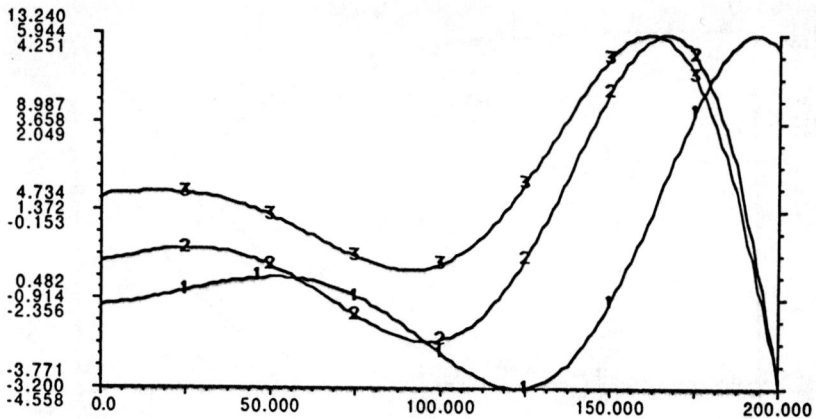

Figure 4 Fluctuations of capital K in a 3-D spatial network with one node as attraction centre and two nodes as growth centres (without a chaotic regime for the attractiveness); y-axis $=K^1$, K^2, K^3 (in descending order from the top), x-axis = time

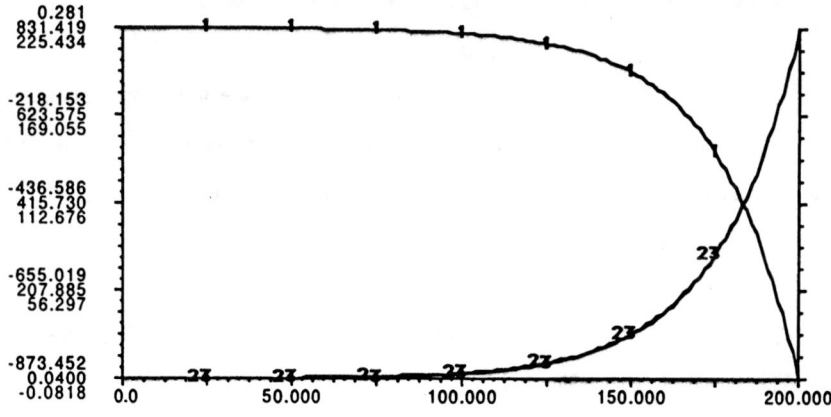

Figure 5 Stable evolution of capital K in a 3-D spatial network with two nodes as attraction centres and one node as a growth centre (under a chaotic regime); y-axis $=K^1$, K^2, K^3 (in descending order from the top), x-axis = time

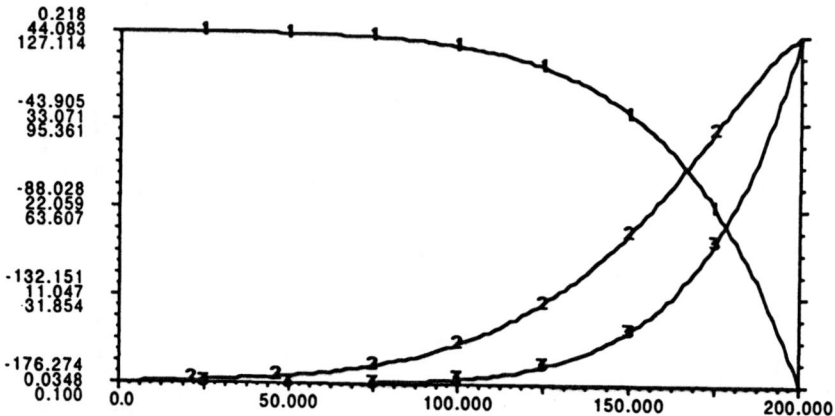

Figure 6 Stable evolution of capital K in a 3-D spatial network, when all nodes are attraction centres; y-axis = K^1, K^2, K^3 (in descending order from the top), x-axis = time

c)   **Centres 1,2, and 3 as attraction centres**

Here we will consider the case in which all nodes are attraction centres and hence competing. According to the Smith hypothesis we get again a stable behaviour for capital, like in the previous case, despite the chaotic regime for the attractiveness (see Figure 6).

After the simple example related to a 3-D spatial network, we will now consider the case of a 4-D spatial network, for both a fully connected system (e.q. a matrix structure) and a partially connected system (e.q. a circular structure).

11.5   STABILITY AND COMPLEXITY IN A 4-D FULLY CONNECTED HIERARCHICAL SYSTEM

Following the approach adopted in the previous section we will now consider different kinds of hierarchy for a 4-D spatial network in which the four centroids have a full connectivity structure (in other words, all the centres are contiguous to each other; see Figure 7):

a) centre 1 is an attraction centre (and centres 2, 3 and 4 are growth centres)
b) centres 1 and 2 are attraction centres (and centres 3 and 4 are growth centres)
c) centres 1, 2 and 3 are attraction centres (and centre 4 is a growth centre)
d) centres 1, 2, 3 and 4 are four attraction (competing) centres.

Analogously to the previous example (i.e., a 3-D network), we will consider here the possibility of chaotic values for the variable A^n.

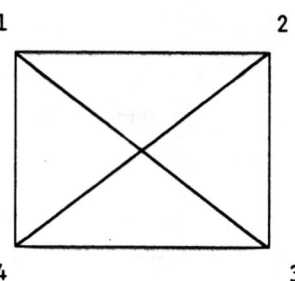

Figure 7   A 4-D fully connected spatial network

a)   Centre 1 as attraction centre

In this simulation experiment we will assume, in addition to the parameter values already adopted in Section 4a, the following parameter values:

$\lambda^4 = 3.9$ $\alpha^4 = 0.05$ $\gamma^4 = 0.7$

$\beta^{14} = 0.04$ $\beta^{24} = 0.01$ $\beta^{34} = 0.03$

$\beta^{41} = -0.01$ $\beta^{42} = 0.02$ $\beta^{43} = 0.04$

$\delta^{14} = 0.03$ $\delta^{24} = 0.015$ $\delta^{34} = 0.03$

$\delta^{41} = -0.09$ $\delta^{42} = 0.05$ $\delta^{43} = 0.01$ $\delta^{44} = 0.01$

$\epsilon^{14} = 0.02$ $\epsilon^{24} = 0.02$ $\epsilon^{34} = 0.02$

$\epsilon^{41} = -0.05$ $\epsilon^{42} = 0.02$ $\epsilon^{43} = 0.05$ $\epsilon^{44} = 0.03$

$b^4 = 1.2$

with the following initial values for the variables:

$A^4 = 0.1$ $K^4 = 0.08$ $Y^4 = 0.1$

and

$P^4 = 0.3$

It is interesting to observe that again in this case of one attraction centre against all other growth centres, we get a fluctuating pattern - for both chaotic and non-chaotic values - for the attractiveness (see Figure 8).

It is noteworthy that for all other cases 5b, 5c, 5d - analogously to the previous section - we get a stable behaviour (see, for example case 5c in Figure 9).

This interesting result leads immediately to the following observations:

- a fluctuating dynamic pattern - at least in the long run - is not so frequent as one would possibly suppose under a chaotic regime
- the relationship competition/stability in a network is still an open research issue which deserves further attention. In this context it is clear that the spatial competition coefficients $\beta^{nn'}$, $\delta^{nn'}$ and $\epsilon^{nn'}$ play a critical role. A first idea on the importance of spatial contiguity in the network may be offered by examining a time correlogram of spatial autocorrelation coefficients as discussed in Section 2. This topic will be dealt with in Section 11.7 with reference to the fluctuating pattern of capital emerging in case 5a.

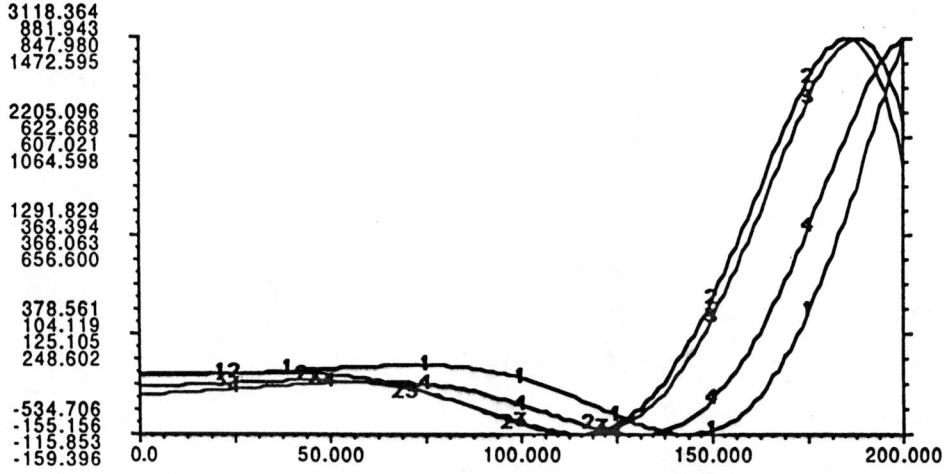

Figure 8 Fluctuating pattern of capital K in a 4-D fully connected network (one node as attraction centre and all others as growth centres), y-axis = K^1, K^2, K^3, K^4 (in descending order from the top), x-axis = time

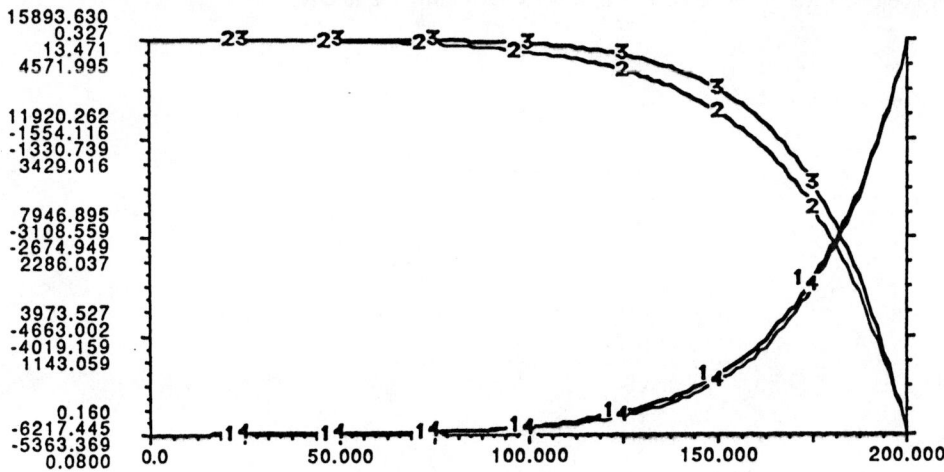

Figure 9 Stable behaviour of capital K in a 4-D fully connected network (three nodes as attraction centres and one node as growth centres); y-axis $=K^1$, K^2, K^3, K^4 (in descending order from the top), x-axis = time

11.6 STABILITY AND COMPLEXITY IN A 4-D PARTIALLY CONNECTED HIERARCHICAL SYSTEM

In the spatial system discussed in Section 11.5 all regions were supposed to be adjacent to one another. In the present section we will assume that the four regions are only pairwise connected, e.g., by forming a belt or a circle. Thus the related centroids are partially connected (see Figure 10). Then again similar simulation experiments can be carried out. In other words, we will again examine four states:

a) centre 1 is an attraction centre
b) centres 1 and 2 are attraction centres
c) centres 1, 2 and 3 are attraction centres
d) centres 1, 2, 3 and 4 are attraction (competing) centres,

the only difference being that a 4-D partially connected network implies no interaction between centres 1 and centre 3 nor between centres 2 and 4.

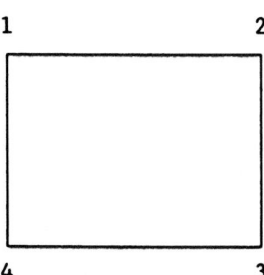

Figure 10   A 4-D partially connected spatial network

a)   **Centre 1 as attraction centre**

In this simulation experiment we will assume the same parameter values as adopted in Section 5a, by assuming, however, no interaction between centres 1 and 3 as well as between centres 2 and 4 viz.:

$$\beta^{13} = \beta^{31} = \delta^{13} = \delta^{31} = \epsilon^{13} = \epsilon^{31} = 0$$

$$\beta^{24} = \beta^{42} = \delta^{24} = \delta^{42} = \epsilon^{24} = \epsilon^{42} = 0$$

It is interesting to observe that also in this case of attraction centre 1 against all other growth centres we get a fluctuating pattern (see Figure 11), which is clearly more visible in the long run (see Figure 12).

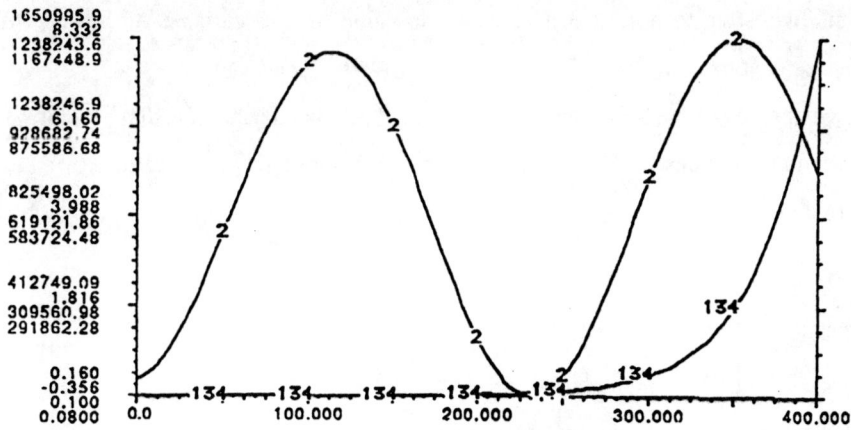

Figure 11 Fluctuating pattern of capital K in a 4-D partially connected network (one node as attraction centre and all others as growth centres); y-axis $= K^1, K^2, K^3, K^4$ (in descending order from the top), x-axis = time

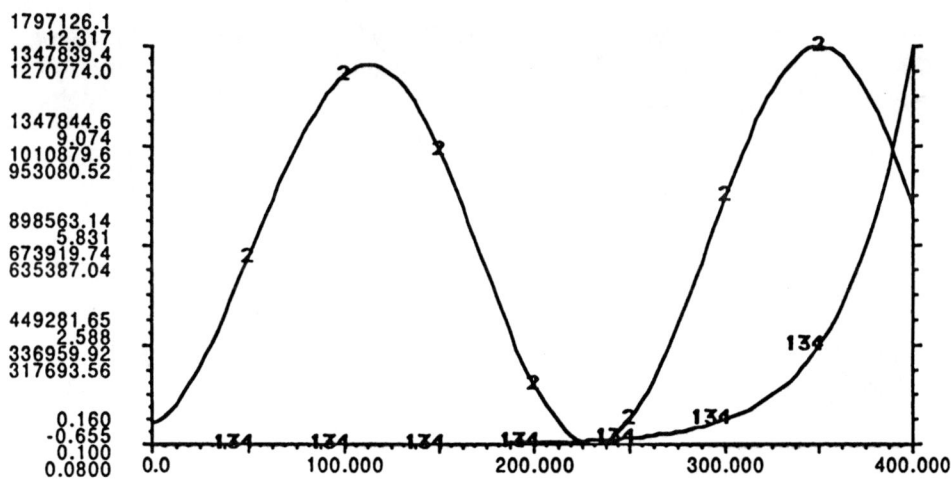

Figure 12 Long run behaviour for the pattern displayed in Figure 11 (under a chaotic regime for the attractiveness); y-axis $= K^1, K^2, K^3, K^4$ (in descending order from the top), x-axis = time

It is interesting to note that a non-chaotic value for the variable A^n has no influence on the behaviour of K^n (see Figure 13 where $A^n = 1.1$).

If we then examine all other cases 6b, 6c, 6d it is easy to see that - analogously to the previous sections - we always get stable behaviour (see, for example, case 6c in Figure 14).

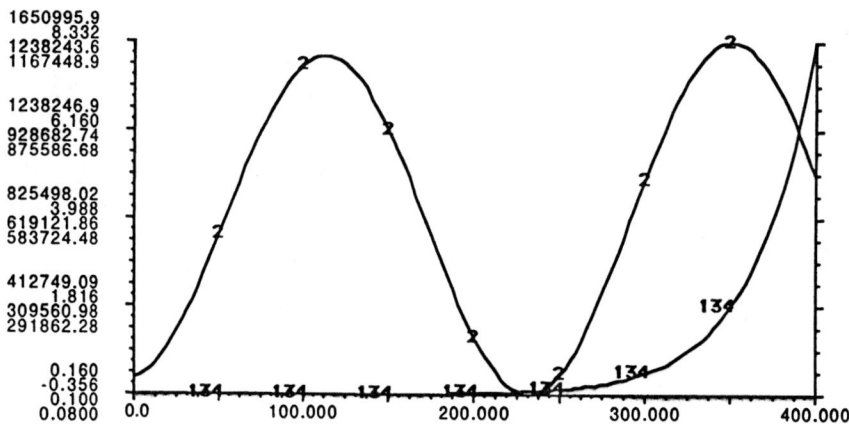

Figure 13 Long run behaviour for capital K (without a chaotic regime for the attractiveness); y-axis = K^1, K^2, K^3, K^4 (in descending order from the top), x-axis = time

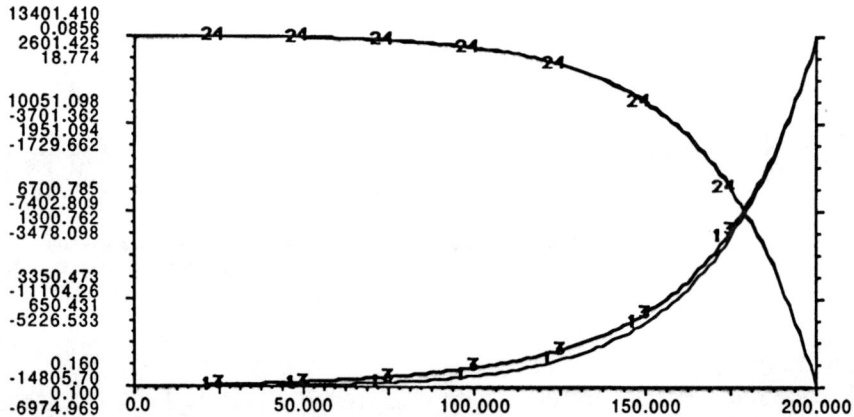

Figure 14 Stable behaviour for capital K in a 4-D partially connected network (three nodes as attraction centres and one node as growth centre); y-axis = K^1, K^2, K^3, K^4 (in descending order from the top), x-axis = time

It is then clear that the results of this section reinforce the conclusions of the previous section, in particular the 'rare' possibility of fluctuating patterns for capital, even after increasing the 'complexity' of the network (number of competitors, number of interactions, etc.). In other words, by increasing the 'complexity - aspect' of the network (number of competitors), we usually get stability for capital. Consequently, in our analysis on network complexity it becomes now relevant to study complexity in a complementary way by investigating the **connectivity pattern**, underlined in Section 11.2, in order to evaluate the spatial effect on capital over time. Thus it is now interesting to examine how slightly different spatial configurations can lead to dissimilar spatial influences.

11.7 SPATIAL CONTIGUITY ANALYSIS FOR A 4-D SPATIAL NETWORK

In this section we aim to examine the time correlogram for the variable K^n with reference to the 4-D spatial network previously analyzed. In particular we will consider the two cases 5a and 6a (a fully connected and a partially connected structure, respectively, with only one node as attraction centre) by evaluating how the spatial influence of surrounding/contiguous centres on the variable 'capital' can vary over time.

In particular for our experiments we will adopt for x_i the simulated time series values for capital obtained in the previous sections.

By adopting now, the M coefficient (11.2) correlated with hypothesis (11.5) it is easy to see that the M coefficient is **constant** for case 5a, since all the centres are connected with each other (see Figure 7). Thus, this case is not interesting for a time correlogram scheme. On the contrary, by considering case 6a (where not all the centres are mutually connected) for the variable K^n in its first 70 time periods simulations, we get a time correlogram of the following type (see Figure 15):

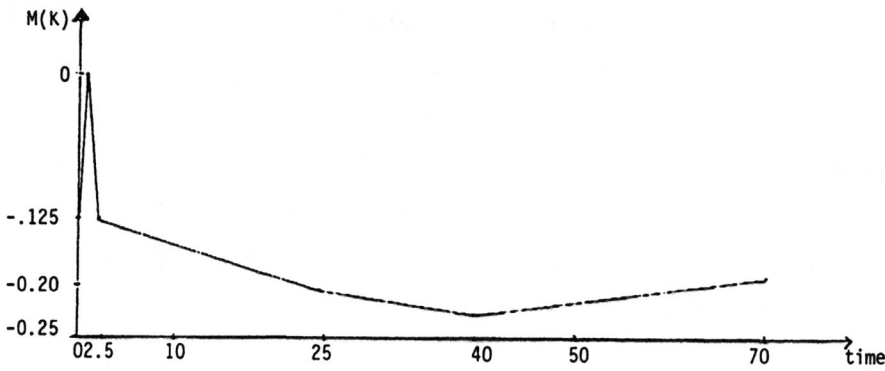

Figure 15 Time correlogram for the Moran coefficient related to capital K in a 4-D partially connected network

From Figure 15 we can see that in this analysis related to the first 70 time periods the Moran coefficient expression in (11.2), (11.3) and (11.5) display values of M > M*, where:

$$M^* = -1/(n-1) \qquad (11.12)$$

is the expected value under the assumption of **no correlation** (see again Cliff and Ord, 1981). Thus the M index in Figure 15 indicates a certain degree of similarity, clustering and smoothness among our centres for the case displayed in Figure 10 (a 4-D partially connected network).

It would also be interesting to verify this spatial influence over a longer period where - as we have shown - the value of K^2 becomes more fluctuating. But obviously this analysis becomes more difficult from the computational viewpoint.

An important conclusion may now be drawn from this experiment. The index of spatial autocorrelation and hence its correlogram depends strongly on the forms of the weights. Thus in a binary form (1,0), for example, the case of totally connected centres with no directional pattern gives no significant spatial autocorrelation over time (a constant value). Obviously, by assuming more appropriate forms for the weight based on distance functions like in (11.6) (or on other functional measures), we can certainly obtain time correlograms reflecting also the weight form (i.e., the strength of the link) in a different way than the usual connectivity matrix.

11.8 CONCLUSIONS AND FUTURE RESEARCH DIRECTIONS

The present study - eventhough in a preliminary stage - has shown how a network can be conceived of as a complex 'space-time' system, whose evolution depends on critical variables/factors which are interrelated in space and time by means of a connectivity structure.

A first analysis related to some complexity aspects (viz. the number of nodes, the number of competitors, the number of connections among nodes) has been carried out for a simple spatial configurations structure (dealing with an interwoven three and four node network), displaying the evolution of three economic variables for the nodes (attractiveness, capital and production). In particular - from the present analysis - the following observations emerge:

- the relationship stability/complexity still deserves a thorough attention, since these first results seem to reinforce the Smith's hypothesis according to which the increase of competing species (in our case the nodes) leads to more stability. Thus this conclusion should be essentially tested for a larger network with a greater number of competing nodes.
- the spatial-contiguity effect is strongly dependent on the connectivity structure as well as on the specification of the weights in the spatial - contiguity formulation.

Thus, also in this context further research is necessary, by considering larger networks with different connections and levels of interactions as well as by formulating more 'functional' specifications for the weights-variables. An interesting research line in this latter respect could be the substitution of an 'information' concept for the 'distance concept' in a spatial analysis, by means of information theory. This new form also allows the description of a spatial network by means of a neural network approach with cellular automata models (see e.g., Fischer, 1992 and White and Engelen, 1993). Consequently also the complexity/(in)stability concept may then be revisited by means of a neural network approach.

ACKNOWLEDGEMENT

The present paper has been developed under funding of the Italian C.N.R. Progetto Finalizzato Trasporti 2 - contract n° 92.01943.PF74

REFERENCES

Anselin, L., 1988, **Spatial Econometrics: Methods and Models**, Kluwer, Boston.

Casti, J., 1979, **Connectivity, Complexity and Catastrophe in Large Scale Systems**, John Wiley, Chichester.

Batten, D., 1987, The Balanced Path of Economic Development: A Fable for Growth Merchant, **Economic Evolution and Structural Adjustment**, (D. Batten, J. Casti and B. Johansson, eds.), Springer Verlag, Berlin, pp. 64-85.

Bennett, R.J. and R.J. Chorley, 1978, **Environmental Systems. Philosophy, Analysis and Control**, Methuen, London.

Cliff, A.D. and J.K. Ord, 1973, **Spatial Autocorrelation**, Pion, London.

Cliff, A.D. and J.K. Ord, 1981, **Spatial Processes**, Pion, London.

Fischer, M.M., 1992, Neural Network Models and Interregional Telephone Traffic, Paper presented at the NECTAR Workshop, Amsterdam.

Geary, R.C., 1954, The Contiguity Ratio and Statistical Mapping, **The Incorporated Statistician**, Vol. 5, pp. 115-145.

Griffith, D.A., 1992, Spatial Regression Analysis on the PC, Research Paper, Dept. of Geography, Syracuse University, Syracuse.

Hordijk, L. and P. Nijkamp, 1977, Dynamic Models of Spatial Autocorrelation, **Environment and Planning**, Vol. 9, pp. 505-519.

May, A., 1976, Simple Mathematical Models with Very Complicated Dynamics, **Nature**, vol. 261, pp. 459-462.

Moran, P.A.P, 1950, Notes on Continuous Stochastic Phenomena, **Biometrika**, Vol. 37, pp. 17-23.

Nijkamp P. and A. Reggiani, 1992a, **Interaction, Evolution and Chaos in Space**, Springer, Berlin.

Nijkamp, P. and A. Reggiani, 1992b, Competition and Ecologically Based Models, Spatial Competition and Ecologically Based Socio-Economic Models, **Socio-Spatial Dynamics**, vol. 3, no. 2, pp. 89-109.

Paelinck, J. and P. Nijkamp, 1975, **Operational Theory and Method in Regional Economics**, Saxon House, Farnborough, Hampshire.

Smith, M.J., 1974, **Models in Ecology**, Cambridge University, Press, London.

White, R. and G. Engelen, 1993, Complex Dynamics and Fractal Urban Form, **Non Linear Evolution of Spatial Economic Systems**, (P. Nijkamp and A. Reggiani, eds.), Springer Verlag, Berlin, (this volume).

CHAPTER 12
DYNAMICAL BASIS OF LARGE DEVIATIONS AND POWER LAWS IN COMPLEX SYSTEMS
G. Nicolis

12.1 INTRODUCTION

One of the most unexpected results that came out of research in the physical and mathematical sciences over the last years is that quite ordinary and perfectly deterministic systems obeying to simple laws, can give rise spontaneously to behaviours of considerable complexity associated with abrupt transitions, a multiplicity of states, or a random-looking evolution to which one refers as deterministic chaos. This suggests that the action of elementary laws over a large number of units constituting a system and over long time periods can result in highly unexpected structures and events characterized by new, *emergent properties* absent at the level of the constituting elements (Nicolis, 1989). The need to devise methods for tackling these phenomena has led to the development of Nonlinear Science, which constitutes currently one of the most active and rapidly growing fields of research (Guckenheimer and Holmes, 1983).

Among the complex systems encountered in our everyday experience economy occupies, undoubtedly, a privileged place. It is therefore natural to inquire whether progress in nonlinear phenomena and complex systems achieved in physical and mathematical sciences may be the starting point of a useful cross-fertilization with economic and social sciences, where complex behavioural patterns arising from cooperative interactions are ubiquitous. The aim of this paper is, first, to draw attention on the fact that certain quantitative *empirical* laws of economics are also present in large classes of complex systems encountered in physical sciences. We shall then inquire whether present day nonlinear dynamics and complex systems research can help identifying generic mechanisms that could lead to these properties. Hopefully, this will in turn give some ideas on how to model them in an adequate manner.

12.2 LARGE DEVIATIONS AND POWER LAWS: PLAUSIBLE DYNAMICAL SCENARIOS

Economic time series are often characterized by the coexistence of two trends : a rather regular long-term trend, reflecting the influence of large scale constraints, and a random looking short term variability (Figure 1a). A most important point to be stressed in this respect is that large deviations from the long term trend are frequently observed (Schiller, 1991), in flagrant opposition with the predominance of small deviations predicted by the classical laws of probability theory (Feller, 1968) such as the central limit theorem and observed experimentally in a great variety of physical systems (Figure 1b).

A second type of behaviour that appears exotic from the standpoint of ordinary physics but is ubiquitous in economic and social sciences are power laws (Montroll and Badger, 1974). Two typical examples are the Pareto distribution and the Zipf law governing such diverse phenomena as the frequency distribution of incomes (Figure 2) or the size of a community plotted against a "rank" parameter varying in the decreasing order of population size. Since such laws are, by definition, scale invariant they constitute the signature of phenomena exhibiting no preferred scale or, equivalently, a wide spectrum of coexisting scales in close interaction. On a more dynamical basis, such "turbulent" systems are expected to possess long term memory and to give rise to slowly decaying correlations in both space and time.

In the remaining of this paper we attempt to identify classes of *deterministic* dynamics capable of producing the types of behaviour summarized in Figures 1a or 2. We shall analyse two scenarios dealing, successively, with low-order dynamical systems and with spatially extended dynamical systems.

12.3 LOW-ORDER DYNAMICAL SYSTEMS

The standard form of the laws governing the evolution of a low-order dynamical system is

$$\frac{d\mathbf{X}}{dt} = \mathbf{F}(\mathbf{X}, \lambda) \tag{12.1}$$

where **X** is the state vector $\mathbf{X} = (X_1,..., X_n)$, n a finite "small" number, **F** a dissipative operator (usually a nonlinear function of **X**) and λ set of control parameters built in the system.

It is by now well-established that equation (12.1) can generate complexity through the *bifurcation* of new branches of solutions following the loss of stability of a certain reference state $\mathbf{X}_s$, as the parameters λ are varied. These bifurcating branches can in turn describe regular or chaotic behaviours.

A. Regular behaviour

In order to reproduce the natural variability summarized in Figure 1 it is necessary to augment equation (12.1) by incorporating the effect of *fluctuations*, which are present in all physical systems. It can be shown that this can be achieved adequately by adding a small-amplitude random force term **f**(t) in the right hand side of (12.1), representing a Gaussian white noise whose variance is related to the macroscopic observables through a law referred to as the fluctuation-dissipation theorem. One thus arrives at

$$\frac{d\mathbf{X}}{dt} = \mathbf{F}(\mathbf{X}, \lambda) + \mathbf{f}(t) \tag{12.2}$$

or, expanding around the reference state $\mathbf{X}_s$,

$$\mathbf{X} = \mathbf{X}_s + \mathbf{x} \tag{12.3}$$

$$\frac{d\mathbf{x}}{dt} = \mathbf{L} \cdot \mathbf{x} + \mathbf{N}(\mathbf{x}) + \mathbf{f}(t) \tag{12.4a}$$

where $\mathbf{L} = (\delta \mathbf{F}/\delta \mathbf{X})_x$ is the linearized operator, $\mathbf{N}(\mathbf{x})$ the nonlinear part, and

$$\langle f_i(t) f_j(t') \rangle = q_{ij} \delta(t - t') \qquad i,j = 1, ..., n \tag{12.4b}$$

A detailed analysis of equations (12.4a) - (12.4b) leads to the following conclusions (Gardiner, 1983, Mansou et al., 1981).

(i) If Reσ_i<0 for all i where {σ_i} are the eigenvalues of **L**, Gaussian fluctuations prevail and correlations decay exponentially. The system behaves as in Figure 1b, provided that the variances {q_{ij}} are small. Typically, this is what happens away from any bifurcation point, when the system possesses a unique asymptotically stable solution.

(ii) If Reσ_i>0 for at least one i = α, fluctuations behave in a non-Gaussian manner and the variance of the state variables **x** tends to be large. This is what happens across a bifurcation point, as depicted in Figure 3. Contrary to case (i) power laws can now appear. For instance at the bifurcation point $\lambda = \lambda_c$ the correlation functions exhibit power law behavior. A similar behavior can occur transiently in a vicinity of λ_c beyond bifurcation (Figure 3c).

B. Chaotic behaviour

Chaotic dynamics provides a mechanism of intrinsically generated randomness in which the variability is comparable to the mean and follows generally a non-Gaussian statistics (Figure 4), much like in the graph of Figure 1a. The abundance of power laws and long-range correlations is a subtler question. In a typical fully-developed chaotic regime none of these is expected to occur. On the other hand it has been shown recently that dynamical systems in the borderline between chaos and periodicity may exhibit long range correlations and, if not a strict power law behavior, at least significant deviations from exponential decay (Gaspard and Wang, 1988, Ebeling and Nicolis, 1992). An interesting such class are *intermittent systems*, in which the evolution consists of long quiescent periods interrupted by briefer turbulent-like bursts of variable duration ignited at times that appear to be randomly distributed.

12.4   SPATIALLY EXTENDED DYNAMICAL SYSTEMS

A typical manifestation of spatial inhomogeneities in extended systems is the occurrence of transport phenomena, whereby a certain property X is transported across the system through a flux $\mathbf{J}_x$ generated by the coupling between spatially adjacent regions. In the absence of sources or sinks X will vary in time only because of transport. The evolution equation replacing equation (12.2) and accounting for the effect of fluctuations is then of the form

$$\frac{\partial X}{\partial t} = - \text{div } \mathbf{J}_x + f_x (\mathbf{r}, t) \qquad (12.5)$$

It can be shown that equation (12.5) generates, inevitably, long range correlations and power laws. For instance, when transport obeys to a Fick-type law $\mathbf{J}_x \approx -D\nabla X$ where D is a positive phenomenological coefficient, the spatial correlation function $C_{xx}(\mathbf{r})$ of $x = X - X_s$ turns out to be (Nicolis and Malek Mansour, 1984)

$$C_{xx}(\mathbf{r}) = <x(\mathbf{r}) \, x(0)> \approx \frac{A}{r^\mu} \qquad (12.6)$$

where $\mu > 0$, A is related to the constraint driving the transport and the bracket denotes an average over all realizations of the random force f_x.

Spatially extended networks of locally coupled elements can also give rise to intrinsically generated variability (in the absence of fluctuating sources) in the form of *spatio-temporal chaos*, a particularly spectacular version of which is *fully developed turbulence*. The energy spectrum E(k) observed in a fluid in the regime of fully developed turbulence is known to exhibit a power law dependence on the wave number k, $E(k) \approx k^{-m}$ where m is close to 5/3, as long as k is in an intermediate range between very small values (long scales) where geometric features and constraints enter explicitly, and large values (small scales) where dissipative effects dominate (McComb, 1990). Although the full theoretical understanding of turbulence is still an open problem, the analysis of the equations of fluid dynamics reproduces the power law behavior and predicts, in addition, corrections to the value of m = 5/3 arising from intermittency. Furthermore while the probability distribution functions of long wavelength excitations (eddies) are approximately Gaussian, the probability distributions of smaller eddies develop increasingly stretched tails (Eggers and Grossman, 1991, Grossmann and Lohse, 1991). Such tails are regarded as one of the most characteristic signatures of intermittency. Similar phenomena arise in simple mathematical models of coupled nonlinear elements like for instance coupled map lattices (Kaneko, 1989). Once again power laws, non-Gaussian behaviour and intermittency appear to be different facets of the same reality.

12.5 DISCUSSION

The analysis carried out in the preceding sections suggests that low-order deterministic dynamical systems giving rise to regular solutions can generate non-Gaussian variability and power law behaviour when subjected to small fluctuations, only in the immediate vicinity of a bifurcation point. In this respect therefore this type of behavior appears to be "non-generic" for such systems.

The situation becomes different when a low-order dynamical system operates in the region of chaotic solutions. Non-Gaussian variability becomes then the rule rather than the exception. Power-law behaviour and long range memory effects can also be achieved if in addition the system belongs to the family of intermittent systems, whose dynamics is in the borderline between deterministic chaos and periodicity. One is tempted to speculate that economic systems, in which memory effects are expected to be important, could be usefully modeled, at least on a global scale, using intermittent chaos as a reference prototype.

Spatially distributed systems of large extension operating in the regime of fully developed spatio-temporal chaos constitute, perhaps, the most typical class combining both non-Gaussian statistics and power law behaviour. Now, in most cases of interest the actors involved in economic activities are also localized in space and interact via the transport of energy, matter and information. It may be expected that these spatial interactions, combined with the inherently nonlinear dynamics of the local elements, give rise to spatio-temporal chaos, thereby generating non-Gaussian and power law behaviors. It would undoubtedly be interesting to conduct modeling studies aiming to substantiate this conjecture.

REFERENCES

Ebeling, W. and G. Nicolis, 1992, **Chaos, Solitons and Fractals**, in press.

Eggers, J. and S. Grossmann, 1991, **Phys. Lett.**, A156, p. 444.

Feller, W., 1968, **An Introduction to Probability Theory and its Applications**, 3rd edition Wiley, New York.

Gardiner, C., 1983, **Handbook of Stochastic Methods**, Springer, Berlin.

Gaspard, P. and X. J. Wang, 1988, **Proceedings National Academy of Science USA**, **85**, p. 4591.

Grossmann, S. and D. Lohse, 1991, **Phys. Rev. Lett.**, 67, p. 445.

Guckenheimer, J. and P. Holmes, 1983, **Nonlinear Oscillations, Dynamical Systems and Bifurcations of Vector Fields**, Springer, Berlin.

Kaneko, K., 1989, **Physica** D34, p. 1.

Malek Mansour, M., C. Vandenbroeck, G. Nicolis and J. W. Turner, 1981, **Ann. Phys.**, 131, p. 283.

McComb, W., 1990, **The Physics of Fluid Turbulence**, Oxford University Press, Oxford.

Montroll, E. and W. Badger, 1974, **Quantitative Aspects of Social Phenomena**, Gordon and Breach, New York.

Nicolis, G., 1989, Physics of far-from-equilibrium systems and self-organization", in **The New Physics**, P. Davies (ed.), Cambridge Univiversity Press.

Nicolis, G. and M. Malek Mansour, 1984, **Phys. Rev.** A29, p. 2845.

Schiller, R.J., 1981, **American Economic Review**, 71, 421. See also La Science du Désordre, **La Recherche**, special issue (May 1991).

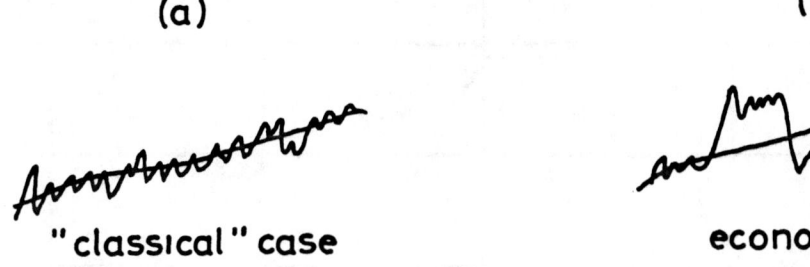

Figure 1   Schematic representation of the variability associated with an economic activity (a), and with a macroscopic observable of a physical system (b).

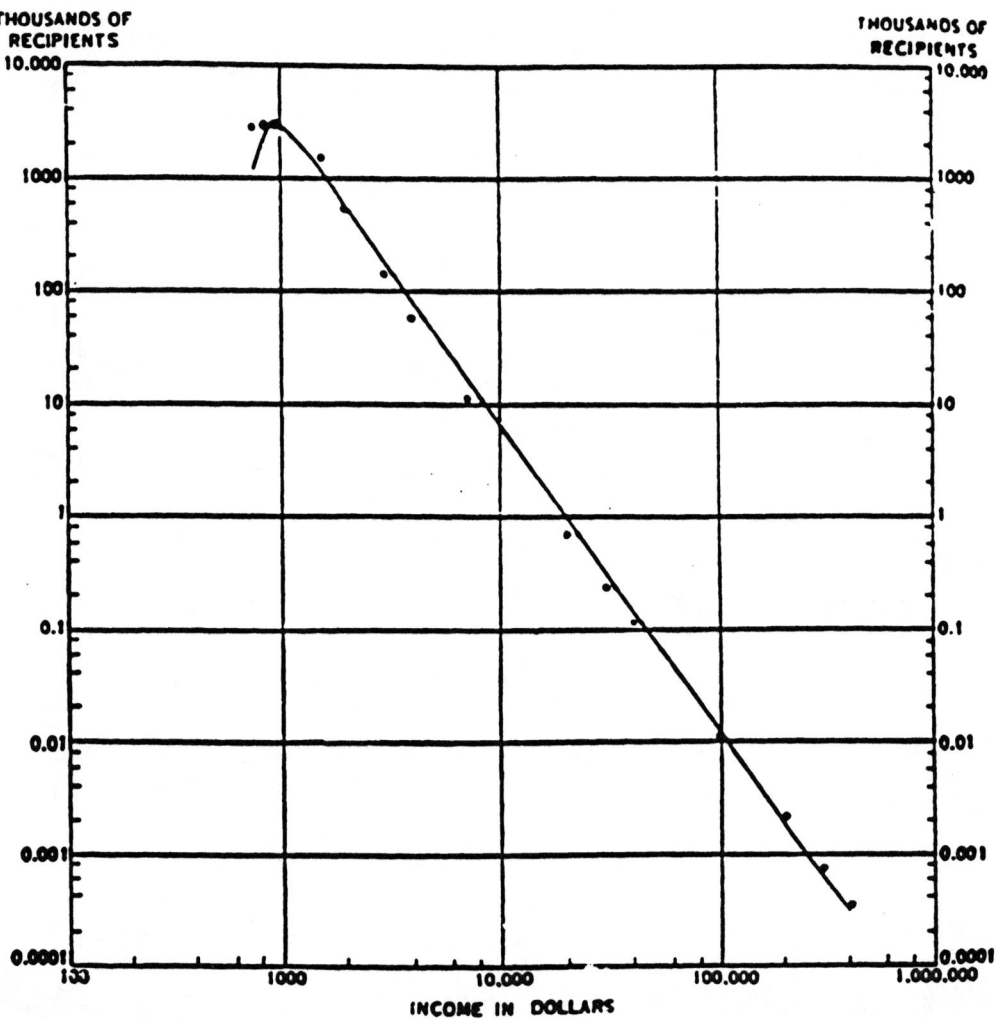

Figure 2   Frequency distribution of incomes in the United States, 1918 (from ref. 5).

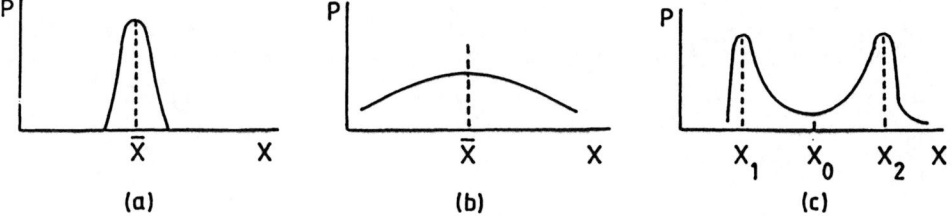

Figure 3   Probability distribution of the variable X inferred from eqs (12.2). As the system crosses the bifurcation point λ_c the probability function changes qualitatively.
(a) $\lambda < \lambda_c$: a unimodal (essentially Gaussian) distribution peaked sharply on a unique state;
(b) $\lambda = \lambda_c$: a flat (non-Gaussian) distribution;
(c) $\lambda > \lambda_c$: a multi-humped distribution whose maxima coincide with the new attractors emerging beyond bifurcation.

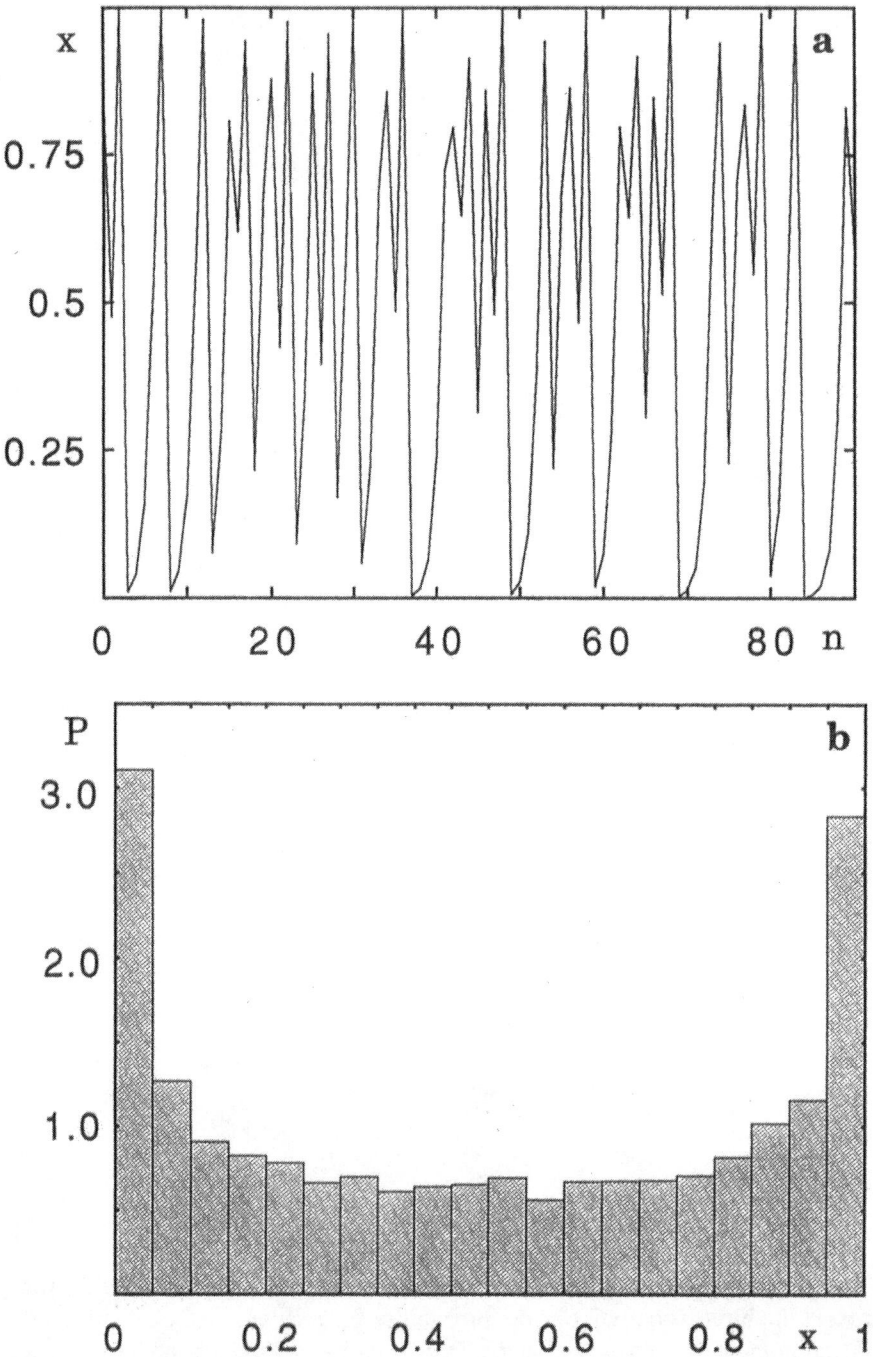

Figure 4 Time series (a) and probability distribution (b) of the variable x generated by the model $x_{n+1} = 4x_n(1 - x_n)$.

CONTRIBUTORS

Peter Allen
International Ecotechnology Research Centre (IERC)
Cranfield Institute of Technology
Bedford MK43 0AL
Uniter Kingdom

Michael Batty
National Center for Geographic Information and Analysis
The State University of New York
105 Wilkeson Quad
Buffalo NY 14261-0001
USA

Guy Engelen
Research Institute for Knowledge Systems
P O Box 463
6200 AL Maastricht
The Netherlands

Gunther Haag
Institute for Theoretical Physics
University of Stuttgart
Pfaffenwaldring 57
D-7000 Stuttgart 80
Germany

M. Hilliges
Institute for Theoretical Physics
University of Stuttgar
Pfaffenwaldring 57
D-7000 Stuttgart 80
Germany

Paul Longley
Department of Geography
University of Bristol
University Road
Bristol BS8 1SS
United Kingdom

Hans-Walter Lorenz
Department of Economics
University of Göttingen
Georg-August Universität
Platz der Göttinger Sieben 3
D-34 Göttingen
Germany

Bill MacMillan
School of Geography
Oxford University
Oxford OX1 3TB
United Kingdom

Grégoire Nicolis
Service de Chimie Physique
Free University of Brussels
Code Postal 231 - Campus Plaine U.L.B.
Bvd du Triomphe
1050 Brussels
Belgium

Peter Nijkamp
Department of Economics
Free University
De Boelelaan 1105
1081 HV Amsterdam
The Netherlands

Jacques Poot
Department of Economics
Victoria University of Wellington
Private Bag
Wellington
New Zealand

Aura Reggiani
Department of Economics
University of Bologna
Piazza Scaravilli, 2
40126 Bologna
Italy

J. Barkley Rosser
James Madison University
Harrisonburg VA 22807
USA

T. Teichmann
Institute for Theoretical Physics
University of Stuttgart
Pfaffenwaldring 57
D-7000 Stuttgart 80
Germany

Roger White
Department of Geography
Memorial University
St. John's Newfoundland
Canada A1B 3X9

Wei-Bin Zhang
Institute for Future Studies
Hagagatan 23B, Box 6799
S-11385 Stockholm
Sweden

The Annals of
Regional Science

An International Journal of Urban, Regional and Environmental Research and Policy

A selection of papers published in Vol. 26

M. M. Fischer, P. Nijkamp: Geographic information systems and spatial analysis

T. Heaps, J. M. Munro, C. S. Wright: A location model of grain production and transportation

D. S. Dendrinos, J. B. Rosser, Jr.: Fundamental issues in nonlinear urban population dynamic models

M. Greenberg, J. Hughes: The impact of hazardous waste Superfund sites on the value of houses sold in New Jersey

D. Holland, S. C. Cooke: Sources of structural change in the Washington Economy

D. N. Steinnes: Measuring the economic value of water quality

J. Rouwendahl: A note on the efficiency of stochastic market equilibria

M. J. Beckmann: Income and expenditure in a spatial economy

H. Beguin: Cristaller's central place postulates. A commentary

M. Fujita, S. Takahashi: Regional income disparity and fiscal-monetary policy. An interregional macroeconomic model of Japan

Editors: D. F. Batten, T. R. Lakshmanan

Special Issues Editor: P. Nijkamp

Book Review Editor: W. P. Anderson

The **Annals of Regional Science** is a quarterly journal in the interdisciplinary field of regional and urban studies. Its purpose is to promote high quality scholarship on the important theoretical and empirical issues in regional science.
The **Annals** publishes papers which make a new or substantial contribution to the body of knowledge in which the spatial dimension plays a fundamental role, such as regional economics, resource management, location theory, urban and regional planning, transportation and communication, human geography, population distribution, and environmental quality.

The **Annals** particularly seeks thoughtful, carefully written articles which are addressed to the broad audience of regional scientists, not just to the author's peers in a subspecialty. Commissioned articles in the journal will aim to focus on what general insights may be gleaned or what general lessons may be learned from an important line of research.

The journal publishes original research, commissioned articles, survey papers, book reviews, and one special issue per annum.